CNC Machining

by

Richard A. Gizelbach

Publisher
The Goodheart-Willcox Company, Inc.
Tinley Park, Illinois
www.g-w.com

Library of Congress Catalog Card Number 2007046909

ISBN 978-1-59070-790-6

3 4 5 6 7 8 9 – 09 – 22 21 20 19 18 17

The Goodheart-Willcox Company, Inc. Brand Disclaimer: Brand names, company names, and illustrations for products and services included in this text are provided for educational purposes only and do not represent or imply endorsement or recommendation by the author or the publisher.

The Goodheart-Willcox Company, Inc. Safety Notice: The reader is expressly advised to carefully read, understand, and apply all safety precautions and warnings described in this book or that might also be indicated in undertaking the activities and exercises described herein to minimize risk of personal injury or injury to others. Common sense and good judgment should also be exercised and applied to help avoid all potential hazards. The reader should always refer to the appropriate manufacturer's technical information, directions, and recommendations; then proceed with care to follow specific equipment operating instructions. The reader should understand these notices and cautions are not exhaustive.

The publisher makes no warranty or representation whatsoever, either expressed or implied, including but not limited to equipment, procedures, and applications described or referred to herein, their quality, performance, merchantability, or fitness for a particular purpose. The publisher assumes no responsibility for any changes, errors, or omissions in this book. The publisher specifically disclaims any liability whatsoever, including any direct, indirect, incidental, consequential, special, or exemplary damages resulting, in whole or in part, from the reader's use or reliance upon the information, instructions, procedures, warnings, cautions, applications, or other matter contained in this book. The publisher assumes no responsibility for the activities of the reader.

Library of Congress Cataloging-in-Publication Data

Gizelbach, Richard.
   CNC machining / Richard A. Gizelbach.
      p. cm.
   Includes index.
   ISBN: 978-1-59070-790-6
   1. Machining. 2. Machine-tools--Numerical control. I. Title.

TJ1185.G54 2008
621.9'023--dc22   2007046909

Introduction

The number of CNC machines used in manufacturing continues to grow for many reasons. A skilled machine operator can run multiple CNC machines at the same time and produce more parts than a machinist using a manual mill and lathe. An experienced machine programmer can instruct a CNC machine to cut material away from a block of steel quickly and accurately. Using CNC machines allows these workers to reproduce parts to the same degree of accuracy throughout the day.

Just as the manufacturing industry has recognized the importance of CNC machining, high schools, colleges, and vocational schools have recognized the need for well-trained operators and programmers. This book is designed to help the student understand the processes, tools, programs, and machines used by successful machine operators and programmers. Although there is a variety of machines and CNC controllers, *CNC Machining* focuses on the G- and M- codes that remain the same from machine to machine. These codes are explained, illustrated, and used in programs found throughout this text. It is suggested that these program examples be inputted into the schools' machine controllers and the parts be produced after necessary editing is performed to conform to the programming format being used with that machine. The programs in *CNC Machining* are written in long form to enable the student to become familiar with these codes. In industry practice, these programs may be shortened extensively to increase efficiency or follow a company format.

This text covers 3-axis CNC machining centers and 2-axis turning centers. However, the same principles are also applied to machines with many more axes and highly sophisticated setups. You should explore the capabilities of these machines by attending trade shows, visiting manufacturer's Web sites, and contacting organizations and associations that support CNC machine operators and programmers.

Read *CNC Machining* thoroughly, edit and run the programs shown in the book, and research the various machines and controllers available, and you will be on the path to a successful career in CNC machining.

4

Brief Contents

Contents

About the Author

Richard A. Gizelbach started his career as a machinist using standard milling machines and lathes. Richard obtained a Bachelor of Science and a Master of Science degree in vocational education from Stout University. He followed this achievement with an EdS degree in Administrative Leadership from the University of Wisconsin-Milwaukee.

Richard's teaching career started with machine shop, CNC, and related courses at Gateway College in Kenosha and Racine counties in Wisconsin. During these teaching years, he conducted customized in-plant training courses in various technical courses for more than 25 firms in the surrounding area, including developing several courses and curriculum for Gateway College and Milwaukee Area Technical College.

After 33 years of teaching, Richard continues to maintain an active schedule by conducting seminars in crane operation and safety, crane inspection, train-the-trainer, and rigging classes for Morris Material Handling.

Your training as a CNC machinist will ultimately pay off here, at the inspection table, where your skills will be judged by the accuracy of the parts you create. (Tibor Machine Products)

N10G20G99G40
N20G96S800M3
N30G50S4000
N40T0100M8
N50G00X3.35Z1.25T0101
N60G01X3.25F.002
N70G04X0.5
N80X3.35F.05
N90G00X5.0Z0T0101
01111
N10G20G99G40
N20G96S800M3
N30G50S4000
N40T0100M8
N50G00X3.35Z1.25T0101
N60G01X3.25F.002
N70G04X0.5
N80X3.35F.05

Chapter 1
Numerical Control and CNC

Objectives

Information in this chapter will enable you to:

- Define the term *numerical control*.
- Describe the origin of numerical control.
- List several advantages of using CNC machine tools.
- List the duties of a programmer.
- List several machine components found on a CNC machine.

Technical Terms

ball screw
binary-coded decimal
    system (BCD)
binary system
closed-loop system
computer numerical
    control (CNC)
computer-aided
    manufacturing
    (CAM)

decimal system
drive motor
electrical numerical
    integrator and
    calculator (ENIAC)
machine body
machine control unit
    (MCU)
machine spindle

numerical control
    (NC)
numerical control
    programming
open-loop system
servomotors
stepper motors
tool changer

Numerical Control

Numerical control (NC) is the operation of a machine tool using a series of sequenced instructions. These instructions consist of alphanumeric characters (letters and numbers). The characters are interpreted (understood) by a computer or *machine control unit (MCU)* and transmitted as electrical impulses. The impulses govern the operation of motors and controls performing various machining functions.

In 1945, engineers Dr. John Mauchly and Dr. J. Presper Eckert created the world's first digital electronic computer at the University of Pennsylvania's Moore School of Engineering. The computer was called an *electrical numerical integrator and calculator,* or *ENIAC.* It was the result of an effort to speed up the scientific and engineering calculations needed during World War II. ENIAC used vacuum tubes and was difficult to program, but the development of this first computer had a dramatic impact upon the industrial society.

The concept of numerical control (NC) came about in the late 1940s, through the efforts of John Parsons of Traverse City, Michigan. While working on a contract to produce helicopter rotor blades, Parsons realized the inaccuracies of using manual methods of production. He devised a method of calculating the coordinate points throughout the entire airfoil surface. Coordinate tables of the contour points were developed, based upon the early model of punched card tabulating equipment. The tabulating equipment was used to generate a table of points spaced no further than 0.5% of the chord length from the previous point. Two Bridgeport milling machine operators, each using a handwheel in X or Y axis, were able to machine patterns more accurately than existing templates used to check contours.

On June 15, 1949, the United States Air Force awarded a contract to Parsons to develop a control system that would govern the movement of the machine to each of the automatically calculated points. In October of that year, Parsons subcontracted with the Massachusetts Institute of Technology to develop servomotors for the machine tool.

The result, in 1952, was a vertical-spindle Cincinnati Hydrotel 3-axis milling machine controlled by vacuum tubes and using straight binary coded tape to provide machine instruction. Simultaneous movement was successful in three axes.

Refinements in development continued throughout the 1960s and 1970s with the number of manufacturers and machines steadily growing. It was not until the late 1970s, however, that computer-based machines were widely used.

Today, computers are less expensive and more powerful, resulting in greater use in machine shops of all sizes. New designs and features have led to multiaxis machines that can perform the most complex of operations to highly accurate standards in less time and at lower cost.

Computer Numerical Control

Computer numerical control (CNC) is the process by which a computer controls the operation of a machine tool. CNC controllers direct the precise movement of a tool along several axes, regulate spindle speeds and cutting feeds, and perform tool changes and various on-off features such as coolant control.

The increasing use of CNC machines is a continuing trend in manufacturing, due to the growth in demand of production output and quality improvement. Because of better engineering designs, advanced tooling, simplified setups, and larger tooling capacities, CNC machines have become faster, more reliable, and more efficient. The functions CNC machines perform make it easier and more profitable for companies to compete in the marketplace.

Advantages of CNC

CNC offers many benefits to manufacturing. These benefits include:

- **Improved safety.** The operator is not as closely associated with the machining process. Guards are put in place to protect the operator. The machine controller is usually located in a safe position, protecting the operator from moving machine parts and the cutting tool.

- **Greater flexibility.** With part programs designed to perform many specialized operations, a wide variety of machining operations can be employed.

- **Greater accuracy.** Removal of the operator from controlling the machine movement has led to greater part accuracy and the ability to hold closer tolerances.

- **Reduced parts inventories.** Reductions in setup time, increased speed of production, and the use of stored part programs eliminate the need to carry a large inventory, because parts can be produced in a shorter amount of time.

- **Reduced lead time.** Stored programs and fewer fixtures allow jobs to be more quickly performed.

- **Lower cutting-tool costs.** Specially ground form tools no longer must be used because the machine can generate a tool path to conform to that tool's shape.

- **Lower fixture costs.** Jobs that once required several fixtures to produce a part now have fewer fixtures; in many cases, only one.

- **Increased cutting tool life.** Speeds and feeds are optimized with CNC machining.

- **Increased production.** Parts can be manufactured faster because there is less interaction between the operator and the machine.

- **Reduced storage space.** Fewer fixtures and fewer parts in storage increases space available for other uses.

- **Complex parts.** The ability of the machine to follow computer-generated tool paths makes it easier to produce complex parts.

- **Less inspection time.** Inspections per order are reduced due to greater accuracy and consistent quality.

- **Fewer machines.** The ability of CNC equipment to perform a variety of operations eliminates the need for many conventional machines.
- **Reduced machining time.** High rapid-positioning feeds and optimized cutting feeds, along with less part handling, greatly reduces machining time.
- **Scrap reduction.** High machining accuracy and reduced human errors lead to less scrap.
- **Easier setups.** Less concern over travel limits and complicated setups makes setups fewer and easier.
- **Better production scheduling.** Scheduling is more accurate because of the controlled machining time for manufactured parts.
- **Cost savings.** All of the listed advantages contribute to an overall cost savings.

Numerical Control Programming

Numerical control programming is the process of combining print information, tooling information, setup information, and speeds and feeds with a sequence of operations. This information is converted into an alphanumeric code that can be interpreted by an MCU. Programmers must interpret part prints, perform math calculations, prepare tooling and setup sheets, write the program in code, and verify and edit the program. The program determines the machine selection, tooling selection, workholding selection, and machining sequence.

Today, the vast majority of programming uses computer-assisted programming software called *computer-aided manufacturing (CAM)* software. This software allows the programmer to create part geometry, select and create tooling, determine machining operations, and create tool paths. The software then writes a G-code program using machine processors for different control brands. G-codes will be covered in Chapter 8.

Decimal and Binary Systems

Most people are familiar with the *decimal system*, which uses base 10 or the power of ten to indicate a numerical value. Computers and MCUs, however, recognize numerical values in the *binary system*, also known as the *base 2 system*.

Decimal System

Numbers in the decimal (also called *Arabic*) system involve the use of digits 0 through 9. The decimal system is comprised on powers of the base number 10. Each place value is ten times larger than the place value directly to its right. For example, the number 17 is made up of 1 tens and 7 ones. See **Figure 1-1.**

Breaking down the whole number 55,319 shows the structure of the decimal system. See **Figure 1-2A**. Remember, any number with an exponent of 0 equals 1, therefore $10^0 = 1$.

Numbers that are less than 1 follow the same principles as whole numbers. Therefore, 10^{-1} equals 0.1, 10^{-2} equals 0.01, 10^{-3} equals 0.001, and 10^{-4} equals 0.0001. A breakdown of the number 0.55319 shows the structure of the decimal system involving a number less than 1. See **Figure 1-2B**.

Binary System

The binary system uses only the digits zero (0) and one (1). This system is basically structured the same as the decimal system. However, the binary system is constructed on the powers of the base 2 and not base 10. This makes each place value twice as large as the place value directly to its right. See **Figure 1-3**. In the binary system, the columns (places) are worth 1, 2, 4, 8, 16, 32, 64, 128, 256, etc.

Number	Ten-thousands	Thousands	Hundreds	Tens	Ones
17				1	7
417			4	1	7
4417		4	4	1	7
64,417	6	4	4	1	7

Figure 1-1. Place value in the decimal system breaks down a number into ones, tens, hundreds, thousands, ten thousands, and so on.

5	5	3	1	9	
$10^4 = 10000$	$10^3 = 1000$	$10^2 = 100$	$10^1 = 10$	$10^0 = 1$	
5×10^4	5×10^3	3×10^2	1×10^1	9×10^0	
50,000 +	5,000 +	300 +	10 +	9 =	**55319**

A

.5	5	3	1	9	
$10^{-1} = .1$	$10^{-2} = .01$	$10^{-3} = .001$	$10^{-4} = .0001$	$10^{-5} = .00001$	
5×10^{-1}	5×10^{-2}	3×10^{-3}	1×10^{-4}	9×10^{-5}	
0.5 +	0.05 +	0.003 +	0.0001 +	0.00009 =	**0.55319**

B

Figure 1-2. Breaking down a number into exponent values. A—The whole number 55,319. B—The decimal number 0.55319.

Binary Number Structure												
2^6	2^5	2^4	2^3	2^2	2^1	2^0	2^{-1}	2^{-2}	2^{-3}	2^{-4}	2^{-5}	2^{-6}
2×2×2×2×2×2	2×2×2×2×2	2×2×2×2	2×2×2	2×2	2	1	$\frac{1}{2^1}$	$\frac{1}{2^2}$	$\frac{1}{2^3}$	$\frac{1}{2^4}$	$\frac{1}{2^5}$	$\frac{1}{2^6}$
64	32	16	8	4	2	1	0.5	.25	.125	.0625	.03125	.015625

Figure 1-3. This chart shows the breakdown of a binary number into the powers of 2 along with its values. Both positive and negative exponents are listed.

The negative exponents in the chart in Figure 1-3 are used for numbers less than 1. Since $2^{-3} = .125$, a one in the –3 position equals .125.

$$2^{-3} = \frac{1}{2^3}$$

$$= \frac{1}{2 \times 2 \times 2}$$

$$= \frac{1}{8}$$

$$= 0.125$$

Numbers are designated as binary numbers by inserting a subscript 2 next to the number. For example, 10_2, 11_2, and 1101_2 are binary numbers. Numbers in the decimal system are understood to be in the decimal system, and do not require a subscript of 10 unless needed for clarity. **Figure 1-4** compares the two systems. The following shows the value of a number (162) as broken down in both the decimal and binary systems.

Decimal

$$1 \times 10^2 = 100$$
$$6 \times 10^1 = 60$$
$$2 \times 10^0 = 2$$
$$= 162$$

Binary

$$1 \times 2^7 = 128$$
$$0 \times 2^6 = 0$$
$$1 \times 2^5 = 32$$
$$0 \times 2^4 = 0$$
$$0 \times 2^3 = 0$$
$$0 \times 2^2 = 0$$
$$1 \times 2^1 = 2$$
$$0 \times 2^0 = 0$$

10100010   = 162

Binary Number Structure					
Decimal Numbers	**Binary Numbers**	**Powers of 2**	**Decimal Numbers**	**Binary Numbers**	**Powers of 2**
0	0		18	10010	
1	1	2^0	19	10011	
2	10	2^1	20	10100	
3	11		21	10101	
4	100	2^2	22	10110	
5	101		23	10111	
6	110		24	11000	
7	111		25	11001	
8	1000	2^3	26	11010	
9	1001		27	11011	
10	1010		28	11100	
11	1011		29	11101	
12	1100		30	11110	
13	1101		31	11111	
14	1110		32	100000	2^5
15	1111		64	1000000	2^6
16	10000	2^4	128	10000000	2^7
17	10001		256	100000000	2^8

Figure 1-4. A comparison of decimal (Arabic) numbers and their equivalents in binary format and powers of 2.

Converting binary to decimal

To convert a number from binary to decimal, simply write it in expanded notation. For example, the binary number 101101_2 can be rewritten in expanded notation as $1 \times 32 + 0 \times 16 + 1 \times 8 + 1 \times 4 + 0 \times 2 + 1 \times 1$. By simplifying this expression, you can see that the binary number 101101 is equal to $32 + 0 + 8 + 4 + 0 + 1$, or the decimal number 45.

Converting decimal to binary

To convert decimal numbers to binary numbers, follow the procedure shown in these examples:

Example: Change 22_{10} to a binary value.

1. Determine the largest power of 2 in 22; $2^4 = 16$ ($2^5 = 32$, which is > 22).

2. Subtract 16 from 22, resulting in 6.

3. Determine the largest power of 2 in 6, $2^2 = 4$ (there is one 2^2).

4. Subtract 4 from 6, resulting in 2.

5. Determine the largest power of 2 in 2; $2^1 = 2$ (there is one 2^1).

6. Subtract 2 from 2, resulting in 0.

7. Since there are no 2^3 and 2^0 values; the place positions for these values must be designated as zeros.

$$\text{Therefore: } 22_{10} = 1(2^4) + 0(2^3) + 1(2^2) + 1(2^1) + 0(2^0)$$

$$\text{In binary form, } 22_{10} = 10110_2$$

Example: Change 14.750_{10} to a binary value

$$2^3 = 8$$
$$14.750 - 8 = 6.750$$
$$2^2 = 4$$
$$6.750 - 4 = 2.750$$
$$2^1 = 2$$
$$2.750 - 2 = 0.750$$
$$2^{-1} = .5$$
$$0.750 - 0.50 = 0.250$$
$$2^{-2} = .25$$
$$0.250 - 0.25 = 0$$

8. There is no 2^0 value, so the place positions for this value must be assigned a zero.

$$\text{Therefore: } 14.750_{10} = 1(2^3) + 1(2^2) + 1(2^1) + 0(2^0) + 1(2^{-1}) + 1(2^{-2})$$

$$\text{In binary form, } 14.750_{10} = 1110.11_2$$

Using the Binary System

The MCU has the ability to recognize values in the standard decimal system and convert them into the binary data. A circuit can be either open (off) or closed (on). A 0 represents open and a 1 represents closed. An adequate number of circuits can be created by combining the 0 and 1 digits to represent any number.

Machines that use perforated tape to load programs using tape readers still exist. With computers dominating the machine tool industry, however, tapes are becoming less common in CNC applications. Tapes have eight channels or vertical rows that may be punched with holes. A hole in the tape designates a closed circuit and represents the digit 1. The absence of a hole designates an open circuit and represents the digit 0. The arrangement of holes in channels (tracks) on the tape determines the place value of the numbers. This is covered in more detail in Chapter 8.

Binary-coded Decimal System

To avoid using a large number of channels (tracks), NC machines use the binary-coded decimal system rather than the binary system itself.

In a straight binary system, a large number of places would be required to represent most numbers. For example, the decimal number 2175, when converted to the binary system is 100001111111. This requires 12 channels. Coding this number on an NC tape would take the form as shown in **Figure 1-5**.

The *binary-coded decimal system (BCD)* uses the decimal system for place locations but converts each digit of the decimal number into a binary number. Using the binary-coded decimal system requires fewer channels, since a row is used to designate a single numerical value as well as alphabetic letters and various symbols. Only four channels are needed to represent the digits from 1 thru 9. Channel 1 has a numerical value of 1, channel 2 has a numerical value of 2, channel 3 has a numerical value of 4, and channel 4 has a numerical value of 8. Combining additional channels with these numerical channels permits the loading of letters and symbols. **Figure 1-6** is a table showing EIA (Electronic Industries Association) and ASCII (American Standard Code for Information Interchange) tape codes.

CNC Machine Construction

The components of a CNC machine can be classified into several units. These include machine body (frame), machine spindles, drive motors, ball screws, tool changers, and computer (MCU).

Machine Body

The *machine body*, or frame, consists of a bed, saddle, column, and table. See **Figure 1-7**. The bed provides the base for all other frame components. It has hardened ways to support the saddle that provides Y axis movement. The table is mounted on the saddle and provides X axis movement. The

Channels	1	2	3	4	5	6	7	8	9	10	11	12	
	2^{11}	2^{10}	2^9	2^8	2^7	2^6	2^5	2^4	2^3	2^2	2^1	2^0	
Binary	1	0	0	0	0	1	1	1	1	1	1	1	
Value	2048					64	32	16	8	4	2	1	=2175

Figure 1-5. Values of the channels that have to be punched on a 12-channel tape to equal the numerical value 2175.

Tape Codes

EIA Character	8	7	6	5	4		3	2	1	Description	8	7	6	5	4		3	2	1	ASCII Character
0			•			•				Numerical Values			•	•		•				0
1						•			•	Numerical Values	•		•	•		•			•	1
2						•		•		Numerical Values	•		•	•		•		•		2
3				•		•		•	•	Numerical Values			•	•		•		•	•	3
4						•	•			Numerical Values	•		•	•		•	•			4
5				•		•	•		•	Numerical Values			•	•		•	•		•	5
6				•		•	•	•		Numerical Values			•	•		•	•	•		6
7						•	•	•	•	Numerical Values	•		•	•		•	•	•	•	7
8					•	•				Numerical Values	•		•	•	•	•				8
9				•	•	•			•	Numerical Values			•	•	•	•			•	9
a		•	•			•			•	A Axis Address		•				•			•	A
b		•	•			•		•		B Axis Address		•				•		•		B
c		•	•	•		•		•	•	C Axis Address	•	•				•		•	•	C
d		•	•			•	•			Special Function		•				•	•			D
e		•	•	•		•	•		•	Special Function	•	•				•	•		•	E
f		•	•	•		•	•	•		Feed Rate	•	•				•	•	•		F
g		•	•			•	•	•	•	Preparatory Function		•				•	•	•	•	G
h		•	•		•	•				Dwell Time, Cycle Repetitions		•			•	•				H
i		•	•	•	•	•			•	Locator for X Axis	•	•			•	•			•	I
j		•		•		•			•	Locator for Y Axis	•	•			•	•		•		J
k		•		•		•		•		Locator for Z Axis		•			•	•		•	•	K
l		•				•		•	•	Locator for 4th Axis	•	•			•	•	•			L
m		•		•		•	•			Miscellaneous Function		•			•	•	•		•	M
n		•				•	•		•	Sequence Number		•			•	•	•	•		N

(Continued)

Figure 1-6. A comparison of the channels that may be punched on a tape to show numerical values and characters using EIA code (odd numbers holes) and ASCII code (even number holes). ASCII code is the most frequently used code.

Tape Codes

EIA	EIA	EIA	EIA	EIA	EIA	EIA	EIA	EIA	Description	ASCII	ASCII	ASCII	ASCII	ASCII	ASCII	ASCII	ASCII	ASCII
Char	8	7	6	5	4	·	3	2	1		8	7	6	5	4	·	3	2	1 · **Char**
o			●			•	●	●		Sequence Number and Scan Stop	●	●			●	•	●	●	● · O
p			●	●		•	●	●	●	Special Function		●		●		•			· P
q			●	●	●	•				Special Function	●	●		●		•			● · Q
r			●		●	•			●	Rapid Traverse Change Point (Z Axis)	●	●		●		•		●	· R
s		●		●		•		●		Spindle Speed		●		●		•		●	● · S
t		●				•		●	●	Tool Function	●	●		●		•	●		· T
u		●		●		•	●			Auxiliary Axis Parallel to X Axis		●		●		•	●		● · U
v		●				•	●		●	Auxiliary Axis Parallel to Y Axis		●		●		•	●	●	· V
w		●				•	●	●		Auxiliary Axis Parallel to Z Axis	●	●		●		•	●	●	● · W
x		●		●		•	●	●	●	X Axis Address	●	●		●	●	•			· X
y		●		●	●	•				Y Axis Address		●		●	●	•			● · Y
z		●			●	•			●	Z Axis Address		●		●	●	•		●	· Z
+			●	●	●	•				Plus Sign (Ignored by Control)			●		●	•		●	● · +
−			●			•				Minus Sign			●		●	•	●		● · −
Delete	●	●	●	●	●	•	●	●	●	Step to Next Character (Tape Feed)	●	●	●	●	●	•	●	●	● · Delete
Enter	●					•				End of Block					●	•		●	· Line feed
Stop code				●	●	•		●	●	End of Record (Rewind Shop)	●		●			•	●		● · Percent
/			●	●		•			●	Block Delete/ (Clash)	●		●		●	•	●	●	● · /
										Non-recognized character; included for tape use with ASCII terminal	●				●	•	●		● · Enter

Figure 1-6. *Continued.*

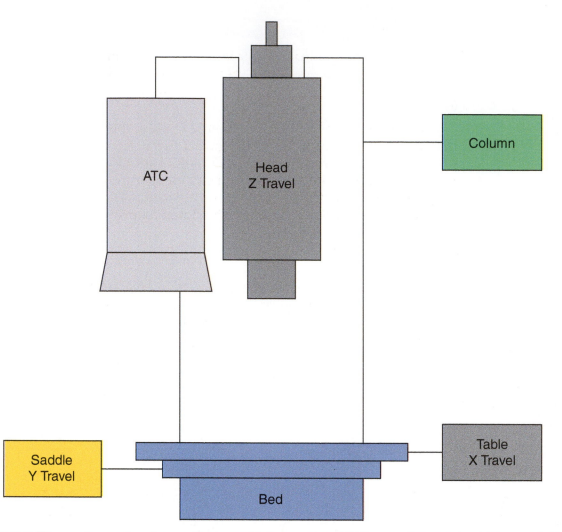

Figure 1-7. The major parts of a vertical machining center. VMCs are the most widely used CNC equipment in industry for machining parts.

column supports the spindle head unit that provides Z axis movement. These components and their movements are described as they relate to a vertical machining center. The X and Z movements are covered in detail in Chapter 2.

Machine Spindles

A *machine spindle* is a device that holds cutting tools (on a CNC mill) or workpieces (on a CNC turning center). See **Figure 1-8**. The spindle rotates clockwise or counterclockwise; on CNC mills, the machine spindle also has the ability to move in and out or up and down.

Spindles have programmable speed ranges. They are driven by electrical motors that are usually direct current (dc), or alternating current (ac). DC motors have infinitely variable speeds. AC motors use a step drive and involve switching from one speed range to another (high to low).

Drive Motors

A *drive motor* controls the machine slide travel on CNC machines. The majority of drive motors are electrical rather than hydraulic.

Stepper motors

Stepper motors convert an electrical pulse provided by an MCU into a finite rotational step. See **Figure 1-9**. For example, the number of pulses sent by the MCU to the stepper motor controls the amount of motor rotation, thus providing slide movement. Stepper motors are used in low-torque situations, with open-loop control systems. These motors are being replaced by servomotors used with the closed-loop system. Stepper motors are also called *stepping motors*.

Servomotors

Direct current *servomotors,* **Figure 1-10**, are variable speed motors that rotate with applied voltage. They drive ball screws and gear mechanisms providing a higher torque output. This output remains constant throughout most of the speed range.

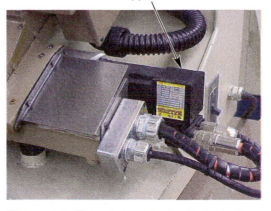

Figure 1-9. This stepper motor is controlling the Z axis on a benchtop South Bend turning center. Open-loop systems use stepper motors.

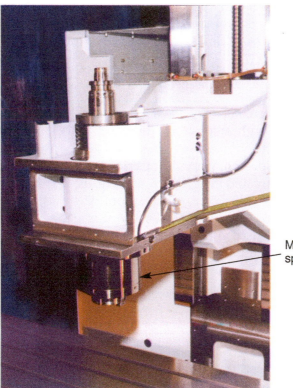

Figure 1-8. The machine spindle on a vertical machining center. On a machining center (also known as a *mill*), the spindle not only rotates, but moves both vertically and horizontally. (Bridgeport Machines, Inc.)

Figure 1-10. This Fanuc ac servomotor is controlling Y axis movement on a horizontal boring and milling machining center. Servomotors use a closed-loop system and can be ac, dc, or hydraulic.

Servometer

Alternating current servomotors are also controlled by regulating voltage that affects speed. These servomotors need less maintenance than dc servomotors.

Hydraulic servomotors are used on large CNC machines because they are more powerful than electrical servomotors. However, they are being replaced with ac servomotors that provide much quieter operation and are nearly as powerful as hydraulic servos.

Positioning control systems

Two types of positioning control systems are used on CNC machines. The *open-loop system* is used with stepper motors and works on the principle of rotary movement of 1.8° for each electrical pulse received. It is an inexpensive system, since it does not provide positional feedback that would require additional hardware and electronics. See **Figure 1-11A**. The *closed-loop system* uses resolvers that continually monitor spindle and table movement and report this data back to the MCU, comparing the current position with the programmed position. See **Figure 1-11B**. Adjustments are made until feedback signals equal transmitted signals. Closed-loop systems, used with servomotors, are very accurate in monitoring slide position and velocity.

Ball Screws

A *ball screw* is a hardened and ground precision lead screw that has a recirculating ball bearing nut used to convert rotating motion to linear motion. See **Figure 1-12**.

Ball screws provide precise positioning and repeatability when converting rotary to linear motion. Recirculating ball screws reduce friction and virtually eliminate backlash that was a problem with acme thread lead screws. Ball screws provide high-speed capability and last longer due to less wear.

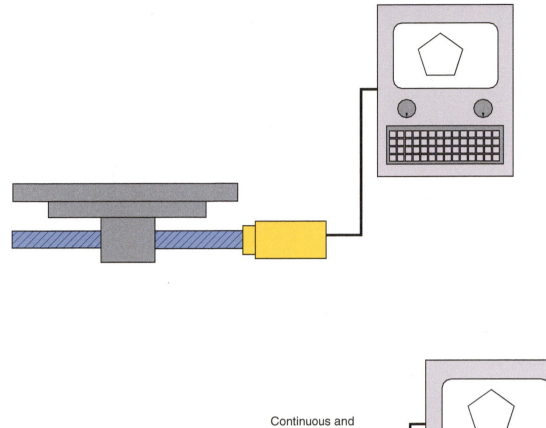

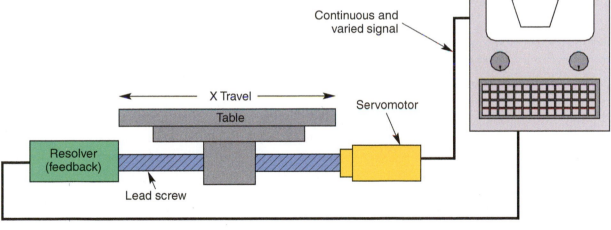

Closed-Loop System
B

Figure 1-11. The machine controller in an open-loop system tells the stepper motor how many times to turn the lead screw. The machine controller in a closed-loop system tells the servomotor how many times to turn the lead screw based on the information (impulses) received from the resolver.

Figure 1-12. A ball screw assembly, consisting of a lead screw and ball bearing nut. Ball screws allow high traversing speeds. (Haas Automation, Inc.)

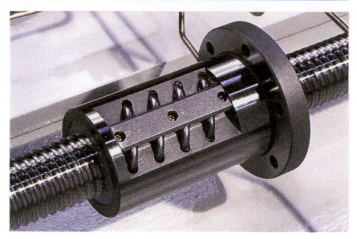

Tool Changers

To manufacture a part, several different tools are typically used. It is, therefore, important to store tools conveniently and provide for rapid tool changes.

A *tool changer* loads and unloads tools to the machine spindle in response to a programmed command given by the machine control unit, **Figure 1-13**. Automatic tool changers with tool magazines store tools in pockets and conduct rapid tool changes without operator intervention. Typically, the changer grips the tool in the spindle, removes it, and replaces it with another tool. This is done in approximately 5 seconds. Most turrets or magazines that hold the tools are bidirectional in rotation and use either random or sequential tool selection. Chapter 4 discusses tool changers in more detail.

Figure 1-13. This CNC machine has a tool magazine that can hold twenty tools. (Haas Automation, Inc.)

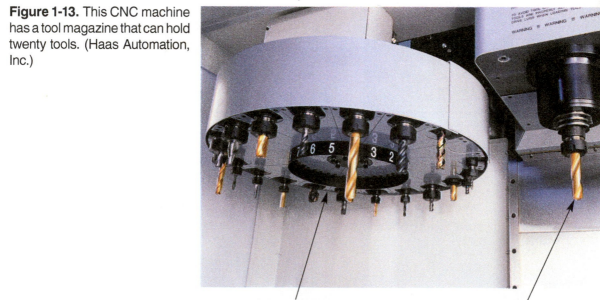

Tool magazine   Current tool

Machine Control Unit (MCU)

The modern machine computer, or machine control unit (MCU), sometimes referred to as the *controller*, contains the memory, computing capability, and circuit switching ability that governs the operation of the CNC machine. See **Figure 1-14**.

The MCU contains ROM (read only memory) that has encoded machine functions built in at the factory. These encoded functions remain in memory even after the computer is shut down. Machine control units vary among controller brands and models, but most have the following features:

- Power on/off buttons
- Emergency stop button
- Feed and speed override knobs
- Load meters
- Manual pulse generator
- Keyboard pad
- Feed hold button
- Cycle start button
- Axis select knobs
- Miscellaneous function switches
- Tool select/clamp switch
- CRT screen

Programs can be stored in memory and retrieved later for processing and editing. Chapter 4 covers more information on data transmission (program loading).

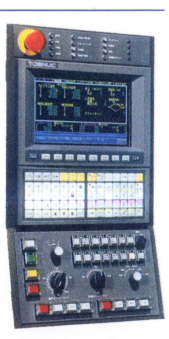

Figure 1-14. The front panel of a typical machine control unit. In addition to controlling machine functions, the unit allows data (programs) to be entered and edited manually using the keyboard. (Toshiba)

Summary

Numerical control is a method of operating a machine tool through the use of coded instructions. These instructions command the machine to perform certain motions and functions. There are more advantages than disadvantages when considering the purchase and use of CNC machines. The first 3-axis milling machine, called the Cincinnati Hydrotel, was operated successfully in 1952.

CNC machines use the binary system in recognizing data. The binary system consists of the digits 0 and 1. The components of a CNC machine can be classified into six units: machine body, machine spindles, drive motors, ball screws, tool changers, and machine control unit.

Chapter Review

Answer the following questions. Write your answers on a separate sheet of paper.

1. List eight advantages of numerical control.
2. What decade began the wide use of computer-based machines?
3. Computer-assisted programming software is sometimes called _____ software.
4. MCUs operate on the binary system (base _____).
5. Tracks are sometimes called _____.
6. List three functions a programmer should perform.
7. Convert the following binary numbers to decimal numbers
    a. $1001_2 =$ _____
    b. $10_2 =$ _____
    c. $11_2 =$ _____
    d. $1.01_2 =$ _____
    e. $11111_2 =$ _____
    f. $10001_2 =$ _____
    g. $10101_2 =$ _____
    h. $1010_2 =$ _____
    i. $10.11_2 =$ _____
    j. $1.1_2 =$ _____
8. Convert the following decimal numbers to binary numbers
    a. $473 =$ _____
    b. $3 =$ _____
    c. $57 =$ _____
    d. $125 =$ _____
    e. $15 =$ _____
    f. $134.125 =$ _____
    g. $0.375 =$ _____
    h. $0.750 =$ _____
    i. $235.0625 =$ _____
    j. $2.5 =$ _____

9. Name two parts of a CNC machine body.
10. Name three components of a CNC machine.
11. List two types of drive motors.
12. What type of positioning control system is used with servomotors?

Activities

1. Other than drills, mills, and turning machines, what other types of machines use CNC? The Internet may be helpful in compiling a list.
2. Describe the origin of numerical control.

These toolholders secure a variety of tools, including drills, carbide inserts, reamers, and face mills.

N10G20G99G40
N20G96S800M3
N30G50S4000
N40T0100M8
N50G00X3.35Z1.25T0101
N60G01X3.25F.002
N70G04X0.5
N80X3.35F.05
N90G00X5.0Z0T0101
O1111
N10G20G99G40
N20G96S800M3
N30G50S4000
N40T0100M8
N50G00X3.35Z1.25T0101
N60G01X3.25F.002
N70G04X0.5
N80X3.35F.05

Chapter 2
Axis and Coordinate System

Objectives

Information in this chapter will enable you to:

- Name the various axes used on machining and turning centers.
- Designate polar coordinates.
- Define the term *absolute measuring system.*
- Define the term *incremental measuring system.*
- Differentiate between the terms *machine zero* and *program zero.*
- Calculate tool positions using Cartesian coordinates on the mill coordinate system.
- Calculate tool positions using Cartesian coordinates on the lathe coordinate system.

Technical Terms

absolute coordinates	incremental	quadrants
absolute measuring	measuring system	signed units
system	machine zero	vector
datum	origin	X axis
Cartesian coordinate	polar angle	Y axis
system	polar axis	Z axis
home position	polar coordinates	zero point
incremental	program zero	zero return position
coordinates		

Cartesian Coordinate System

Cartesian (rectangular) coordinates are the key to understanding the basis of numerical control. Rectangular coordinates are used in everyday life. For example, rectangular coordinates can be seen on city maps showing the layout of streets in a rectangular formation. When we give directions, we often use rectangular coordinates: "the building is three blocks east and two blocks north of here."

The *Cartesian coordinate system* is made of lines or planes that run horizontally and vertically and intersect with each other. These lines or planes are perpendicular (90°) to each other. A third line or plane perpendicular to the other two can be added to this system. With Cartesian coordinates, any point in space can be described in mathematical terms from any other point along three mutually perpendicular (90°) axes.

Any point on a plane can be located by knowing its distance from each of the two intersecting lines. When locating points on a workpiece, two intersecting lines (one horizontal and one vertical) are used. These are at right angles (90°) to each other. These lines are called axes and the point where they cross, or intersect, is called the *origin, datum,* or *zero point.* From the zero point, we may plot points along that line. Points on the plus side of the zero point will have a plus value and points on the minus side of the zero point will have a minus value. The intersection of the lines or planes is designated as point 0 (zero).

The vertical line is called the *Y axis*; the horizontal line is called the *X axis.* This axis designation is applied to machining centers (milling, boring, and drilling machines). When referring to turning centers, the vertical axis is the X axis and the horizontal axis is labeled Z. See **Figure 2-1.**

The construction of CNC machines is based on two or three planes, perpendicular to each other, that correspond to the axes of the Cartesian coordinate system. The three axes for a machining center are designated as X, Y, and Z. See **Figure 2-2.** The two axes for a turning center are designated as X and Z.

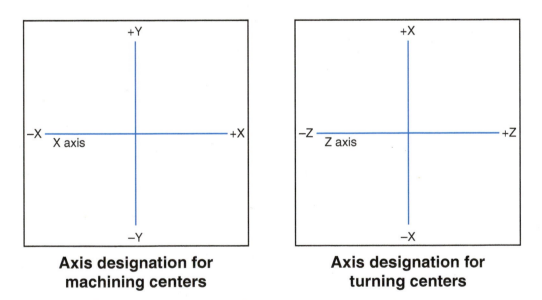

Axis designation for machining centers

Axis designation for turning centers

Figure 2-1. On a CNC machining center, the X axis represents the table and has the greatest travel on the machine. On a CNC turning center, the Z axis is the horizontal axis and represents longitudinal travel.

Two-axis Coordinate System

A two-axis coordinate system consists of four sections called *quadrants*, **Figure 2-3**. The sections are numbered counterclockwise, beginning with the upper-right quadrant.

In the **Figure 2-4**, Point *A* may be described as being located 2 units above the X axis and 3 units to the right of the Y axis. However, to avoid the terms *to the left of, to the right of, below,* or *above,* **signed units** are used. This means numbers have a value sign of plus (+) for positive or minus (–) for negative. Values to the right of axis Y are positive; those to the left are negative. Values above the X axis are positive and below are negative. Point *B* is 3 units below the X axis and 4 units to the right of the Y axis. Therefore, the position of *B* is X= +4, Y= –3. As a rule, the distance to the right or left of the Y axis is given first, and the distance above or below the X axis is given second. Normally, the plus sign (+) does not need to be given. Its absence indicates a positive value.

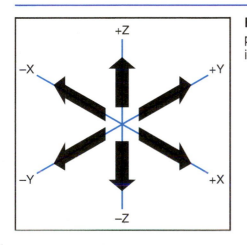

Figure 2-2. Each of the three major axes (X, Y, and Z) is perpendicular (at a 90° angle) to the other two. The arrows indicate the direction of travel.

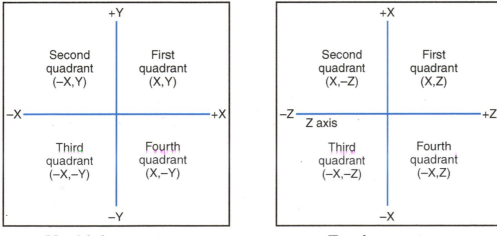

Machining centers   **Turning centers**

Figure 2-3. The four quadrants of a two-axis coordinate system are numbered in a counterclockwise direction. Note the difference in axis designations for machining centers and turning centers.

When the rectangular coordinate system is shifted from the blueprint to the table of a machine tool, the dimensions necessary to machine the workpiece are obtained.

Simple two-axis drilling machines are based on this coordinate system. Usually, the table is designated all positive (plus). This allows all work to be performed in the first quadrant, making all X and Y values positive (+).

For example, **Figure 2-5** represents a drilling machine table with a workpiece located in the upper-right quadrant. The origin, or part zero, is the corner of the machine table or in this case the lower-left corner of the

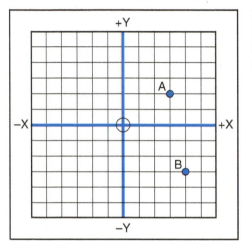

Figure 2-4. Signed (+ or −) units are used to clarify directions in the Cartesian coordinate system. Point A is located in quadrant 1, at coordinates (X+2, Y+3). Point B is located in quadrant 4, at coordinates (X+4, Y−3).

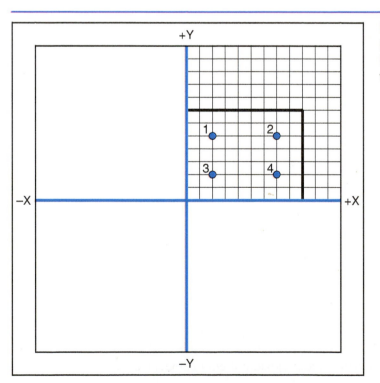

Figure 2-5. A plate on a drilling machine is positioned in quadrant 1, so that all 4 holes will have positive X and Y location values.

workpiece. A series of holes can be drilled in the workpiece by specifying all locations with X-Y coordinate values. Because all the holes are in the first quadrant, all values are positive.

Three-axis Coordinate System

By using a third plane called the *Z axis*, more complicated work can be completed. This plane is perpendicular to the plane comprised of the X and Y axes. See **Figure 2-6**. Using the example of drilling holes in a flat workpiece in the X-Y plane, the hole depth is given as a specific distance along the Z axis. To do this, 3-axis programming is used.

Absolute and Incremental Systems

There are two methods of specifying a location or endpoint to the MCU. The first method is the absolute measuring system; the second is the incremental measuring system. Both use the same distance measuring

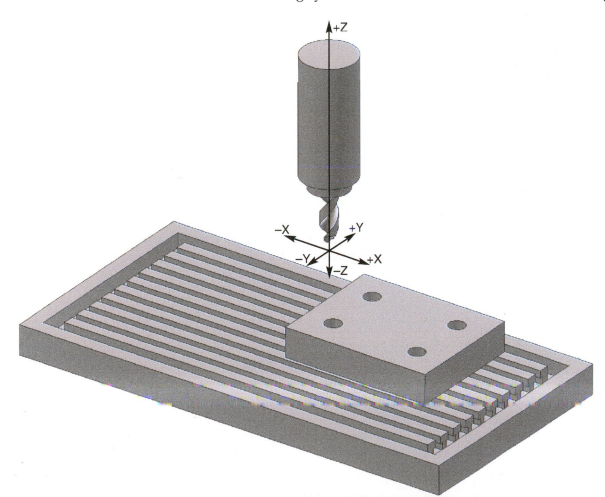

Figure 2-6. The spindle of the drilling machine must travel in a Z direction to drill holes in the plate.

system, whether U.S. customary or metric, but differ in their point of reference. Each method deals with the type of reference coordinate system designated by the programmer or operator.

Absolute Measuring System

The *absolute measuring system* uses measurements from a fixed reference point to specify a point or location, **Figure 2-7**. All the tool locations and movements are given values relating to that original fixed reference point.

Machining center (mill)

Absolute coordinates use the origin point as the reference point. Any point on the Cartesian graph can be plotted accurately by measuring the distance from the origin location to the reference location along the X and Y axes. See **Figure 2-8**. Point A is 3 units along the X axis from the origin and 3 units along the Y axis from the origin. Thus, the point is (X3.0, Y3.0). Point B is 5 units along the X axis from the origin and 6 units along the Y axis from the origin. Thus, the point is (X5.0, Y6.0). Point C is 7.5 units along the X axis from the origin and 7 units along the Y axis from the origin. Thus, the point is (X7.5, Y7.0). **Figure 2-9** is an example of absolute coordinates as applied to a part print.

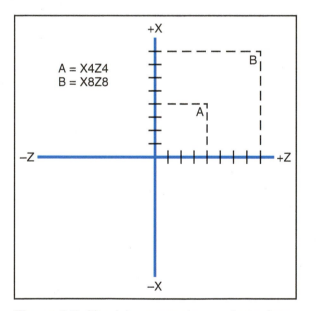

Figure 2-7. Absolute measuring systems base all locations on distance along the axes from a fixed reference point called the *datum* or *origin*.

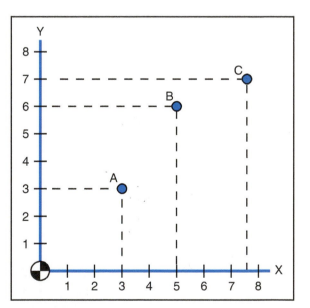

Figure 2-8. Locations A, B, and C are all in the first quadrant. The location of each can be identified by a pair of numbers indicating the distance from the origin along each axis. For example, Point A is three units from the origin along the X axis and three units along the Y axis, which can be expressed as (X3.0, Y3.0).

Position	X	Y	Position	X	Y
1	0	0	12	3.75	3.0
2	0	1.0	13	4.75	3.0
3	.5	1.5	14	5.5	2.25
4	0	2.0	15	5.5	1.25
5	0	2.370	16	4.25	0
6	.63	3.0	17	3.0	0
7	2.25	3.0	18	3.0	.5
8	2.25	2.75	19	1.5	.5
9	2.5	2.5	20	1.5	0
10	3.5	2.5	21	.63	2.370
11	3.75	2.75	22	4.25	1.25

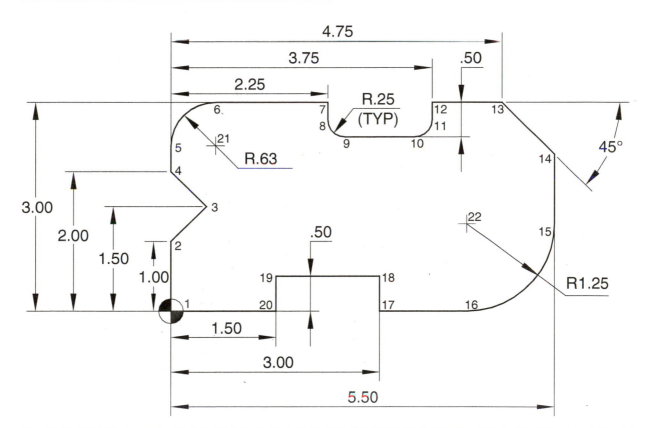

Figure 2-9. This example of datum dimensioning shows the table that lists the values of all the numbered positions shown on the part print.

Turning center

For turning center applications, the absolute coordinates are referenced from a fixed position along the X and Z axes. See **Figure 2-10A**. Position 1 is defined as (X3.0, Z0). To move from Position 1 to Position 2, the location of Position 2 is defined as (X3.0, Z–2.0). Position 3 is defined as (X5.0, Z–3.0), and position 4 is (X5.0, X–6.0). The location of position 5 is defined as (X7.0, Z–6.0). Notice that each position is defined from the origin point, which is referred to as *part zero*. **Figure 2-10B** is a coordinate table listing these values.

Incremental Measuring System

The *incremental measuring system* uses a floating reference point to specify a point or location. Each new tool location and movement uses the last location as a reference point, **Figure 2-11**.

Machining center (mill)

Incremental coordinates use the current position as the reference point for the next move. Incremental moves along the X and Y axes require direction and amount of movement. See **Figure 2-12**. Point A is –6 units along the X axis from origin and –2 units along the Y axis from origin. Point B is +4 units along the X axis from Point A and +8 units along the Y axis from point A. Point C is +6 units along the X axis from Point B and +1 unit along

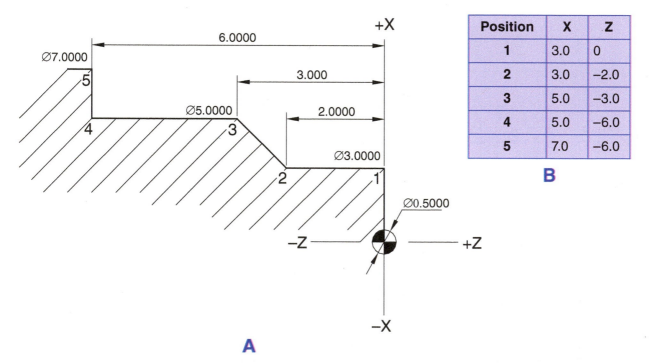

Position	X	Z
1	3.0	0
2	3.0	–2.0
3	5.0	–3.0
4	5.0	–6.0
5	7.0	–6.0

B

A

Figure 2-10. This drawing of a part to be made on a turning center includes a table listing position values.

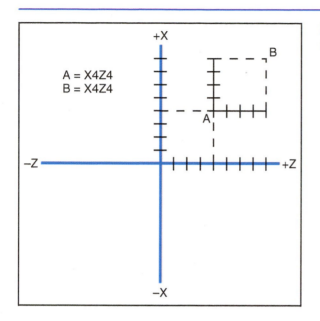

Figure 2-11. Incremental measuring systems use a new origin (datum) for each succeeding movement.

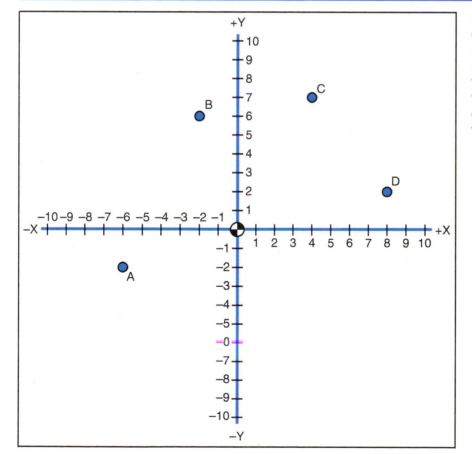

Figure 2-12. Incremental coordinates describe each location from the previous location and require listing the axis involved and the amount (value) and direction of the move to the new location.

the Y axis from Point B. Point D is +4 units along the X axis from Point C and –5 units along the Y axis from Point C. **Figure 2-13** is an example of incremental coordinates as applied to a part print.

Position	X	Y	Position	X	Y
1	0	0	8	2.0	
2		3.5	9		–3.5
3	2.0		10	–.5	–1.0
4		–.75	11	–2.0	
5	1.75		12	–.5	.5
6	.75	.75	13	–.5	–.5
7		1.0	14	–3.0	

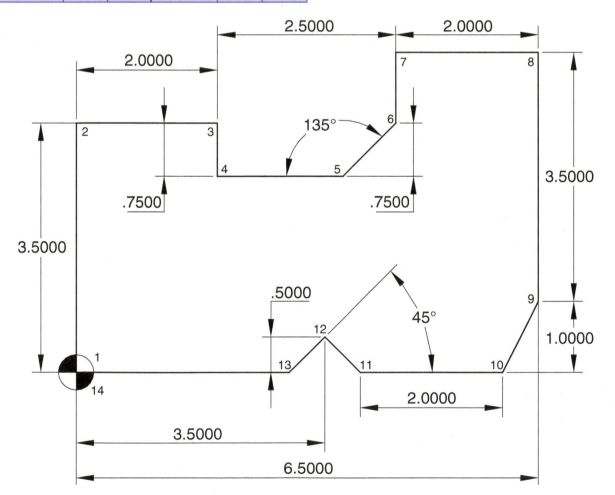

Figure 2-13. The table in this drawing lists incremental moves from Position 1 through Position 14.

Turning center

When incremental coordinates are used with turning centers, the addresses U and W are usually used to replace the addresses X and Z. Therefore, X becomes U and Z becomes W. See **Figure 2-14A**.

From part zero to Position 1, the tool must move a distance of 3″ in the +X (U) direction; no movement is required in the Z (W) direction. Therefore, the move is U3.0. From Position 1 to Position 2 is a distance of 2″ in the –Z (W) direction with no movement required in the X (U) direction. Therefore, the tool move becomes W–2.0. To move the tool from Position 2 to Position 3 requires movement in both axes. Therefore, the move becomes U4.0, W–1.0. **A U move of U4.0 will only move the tool up 2.0″ incrementally on the X axis.** To move the tool from Position 3 to Position 4 requires movement only in the Z (W) axis. Therefore, the move becomes W–3.0. From Position 4 to Position 5, movement is only required in the X (U) axis. Therefore, the move becomes U4.0. The coordinates table of these movements is shown in **Figure 2-14B**.

Absolute vs. Incremental Systems

Figure 2-15A shows an example of absolute coordinates as applied to a part print. **Figure 2-15B** shows the same example using incremental coordinates. Notice the difference between the coordinates tables.

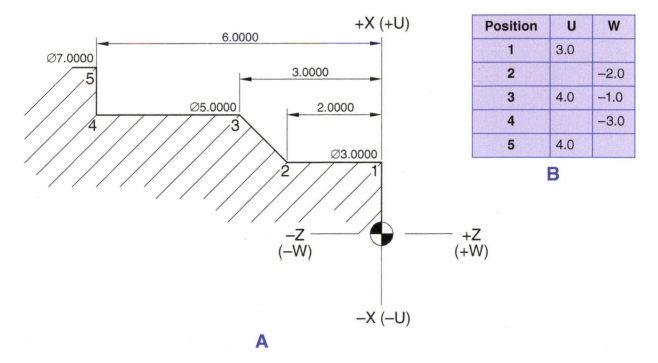

Position	U	W
1	3.0	
2		–2.0
3	4.0	–1.0
4		–3.0
5	4.0	

B

A

Figure 2-14. This turning center drawing contains a table that lists incremental values; substitute the letter U for the X axis and W for the Z axis.

Absolute Coordinates

Position	X	Z	Position	X	Z
1	0	0	7	2.0	−3.25
2	1.0	0	8	3.0	−3.25
3	1.0	−1.25	9	3.0	−4.5
4	1.5	−1.25	10	4.0	−4.5
5	1.5	−2.25	11	4.0	−5.5
6	2.0	−2.25	12	5.0	−5.5

A

Incremental Coordinates

Position	U/(X)	W/(Z)	Position	U/(X)	W/(Z)
1–2	1.0		7–8	1.0	
2–3		−1.25	8–9		−.75
3–4	.5		9–10	1.0	
4–5		−1.0	10–11		−1.0
5–6	.5		11–12	1.0	
6–7		−1.0			

B

Figure 2-15. Compare the values in the absolute and incremental tables that describe the same step shaft.

Polar Coordinates

Some CNC controls use polar coordinates to define the location of points. *Polar coordinates* define the position of a point by its distance and direction from a fixed reference point. Drilling holes in a circular pattern is a good example of the use of polar coordinates.

The distance between points represents a radius value and the direction refers to the angle between the polar (X) axis and the point to be defined. The *polar axis* is a horizontal reference line drawn from the origin out to the edge of a circle. A line drawn from the origin to the desired point is called the *vector*, or radius, **Figure 2-16**. The angle formed by the vector and the polar axis (X) is called the *polar angle*. The angle is positive (+) in the counterclockwise (CC) direction and negative (–) in the clockwise (CW) direction. Polar coordinates are given as (distance, angle), or (R, A). **Figure 2-17** compares polar and Cartesian coordinates.

Machine Zero

Machine zero, or *zero return position,* is a very accurate position along each of the machine's axes that is set by the machine tool builder. Upon machine startup, the machine axes must be sent to the zero return position before operating the machine. This procedure is called *homing the machine* (Fanuc machines) or *cold starting* (Fadal machines).

Machining center

On vertical machining centers, the zero return position is usually the extreme plus travel of each axis. The exception to this is that Fadal machines use the middle of the worktable as the home position. The *home position* is used as a reference point to locate or set the program zero point.

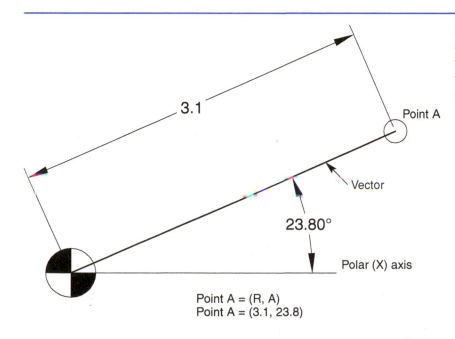

Figure 2-16. Polar coordinates describe a new location from the present location by showing values for the vector and the polar angle.

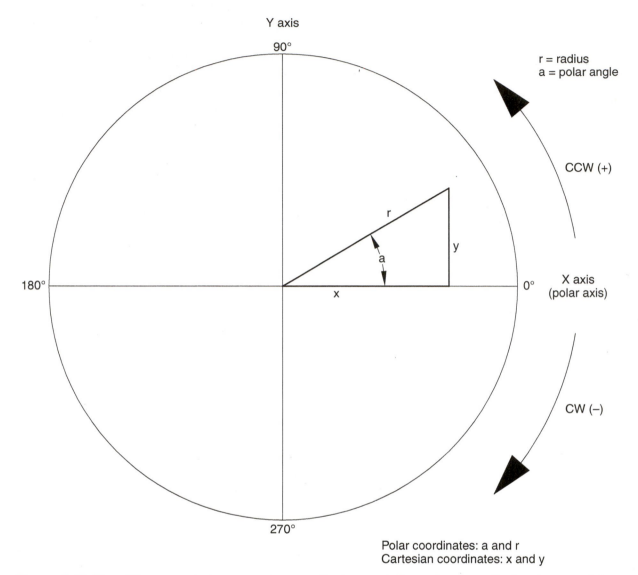

Polar coordinates: a and r
Cartesian coordinates: x and y

Figure 2-17. The differences between polar coordinates and Cartesian coordinates in regard to a location: counterclockwise rotation is positive, clockwise rotation is negative.

Turning center

Turning centers have machine zero located at the extreme plus travel on the X and Z axes. On the universal slant bed turning center, this is the extreme upper-right end of the machine. The setting of program zero is measured from this machine zero location.

Program Zero

Program zero, or part zero, is a floating point (X0,Y0). It can be set at any position inside the machine's grid system limits. It can be located anyplace on the part to establish a start point from which you can calculate the rest of the coordinates. On any drawing, you can usually select the location of the (X0, Y0) point that will be most convenient for calculating the balance of the coordinates. For example, if the part is symmetrical, the center of the part can be used to set the (X0, Y0). This eliminates half the calculations, because the coordinates on both sides of the zero point are the same, except that one set is negative and the other is positive. **Figure 2-18** illustrates two examples of this type of setting.

Machining center

The majority of work performed in a vise has program zero located in the upper-left hand corner of the workpiece. See **Figure 2-19**. This is done so that the fixed jaw and the left end of the vise serve as datum surfaces for locating purposes. Establishing program zero on a workpiece can be called *zero shift* or *work shift*. More information on this process of establishing program zero can be found in Chapter 9.

Turning center

Program zero on a turning center workpiece is usually found on the right end of the workpiece. See **Figure 2-20**. Sometimes the chuck jaws face is used as program zero, but in the majority of cases, the right face of the workpiece is used. However, X0 is always the centerline of the workpiece.

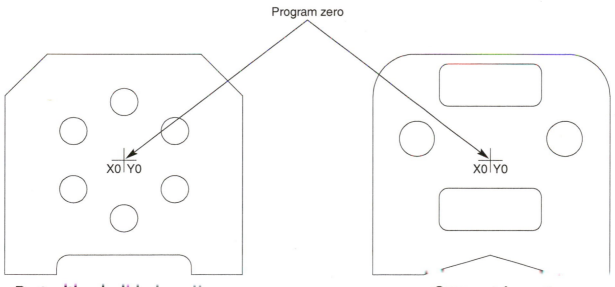

Part with a bolt hole pattern   **Symmetric part**

Figure 2-18. Program zero is positioned to provide easy radial calculations, as in this bolt hole pattern, and symmetric calculations.

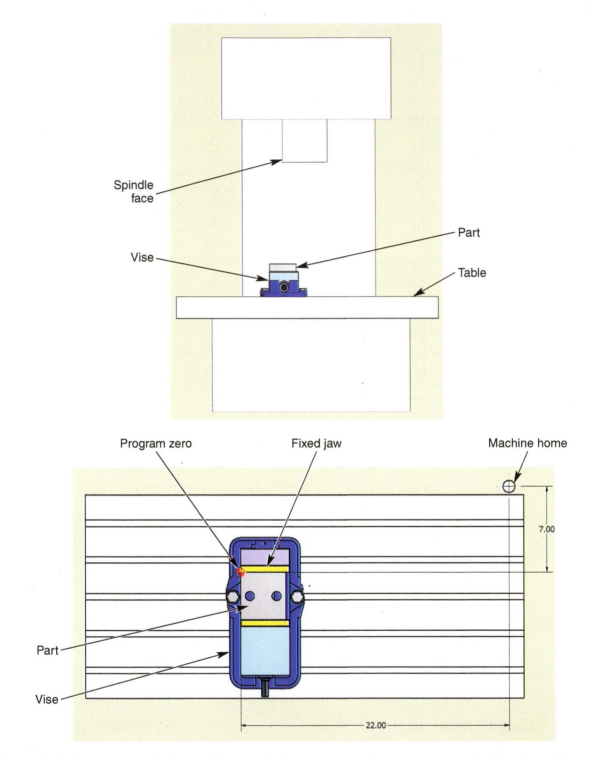

Figure 2-19. Program zero is located in upper-left corner of workpiece because the fixed jaw of the vise remains stationary and zero is not affected by variations in workpiece size.

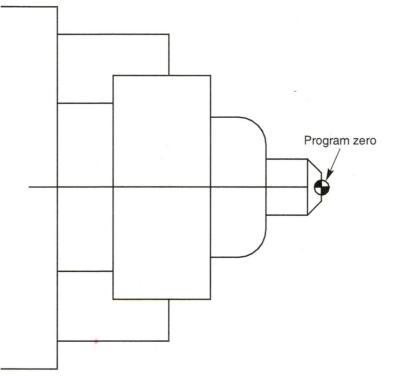

Figure 2-20. On a turning center, program zero is usually at the right end of the workpiece. If excess stock is present at the end, program zero must be set past the excess amount of stock.

Summary

Understanding the Cartesian coordinate system is essential when working with CNC machines, since all tool positions are designated using this system. The Cartesian coordinate system (sometimes called rectangular coordinate system) has four quadrants, which begin in the upper-right position and rotate counterclockwise. Machining centers use X, Y, and Z axes while turning centers use X and Z axes.

The absolute coordinate system uses a fixed zero point while the incremental coordinate system uses a floating zero point. Polar coordinates use distance and direction to define a point. Distance is represented by a radius value and direction refers to the angle between the positive X axis and the point defined.

Machine home is usually the extreme travel in plus directions. Program zero is usually defined from machine home.

Chapter Review

Answer the following questions. Write your answers on a separate sheet of paper.

1. What are the axes values in the third quadrant of the machining center coordinate system?
2. What are the axes values in the second quadrant of the turning center coordinate system?
3. Give two names for the intersection of the axes.

4. Define the term *absolute measuring system.*
5. Define the term *incremental measuring system.*
6. How is a point defined using polar coordinates?
7. Where is program zero usually located on a rectangular part held in a vise?

Activities

1. Use absolute system to identify locations.

Position	X	Y	Position	X	Y
1			11		
2			12		
3			13		
4			14		
5			15		
6			16		
7			17		
8			18		
9			19		
10			20		

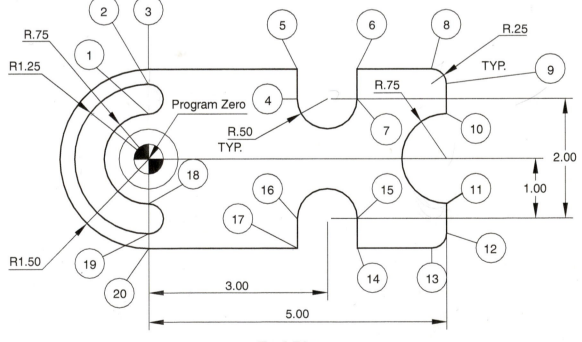

End Plate

2. Use absolute system to identify locations.

Position	X	Y	Position	X	Y
1			11		
2			12		
3			13		
4			14		
5			15		
6			16		
7			17		
8			18		
9			19		
10			20		

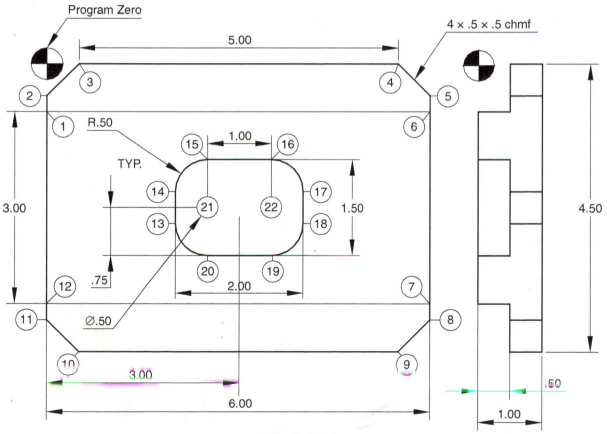

Block Plate

3. Calculate the X and Z coordinate values of the specified locations on the V-shaft using diameter programming and the absolute system. The X values are diameter values. Place the values in the table.

Position	X	Z	Position	X	Z
1			6		
2			7		
3			8		
4			9		
5			10		

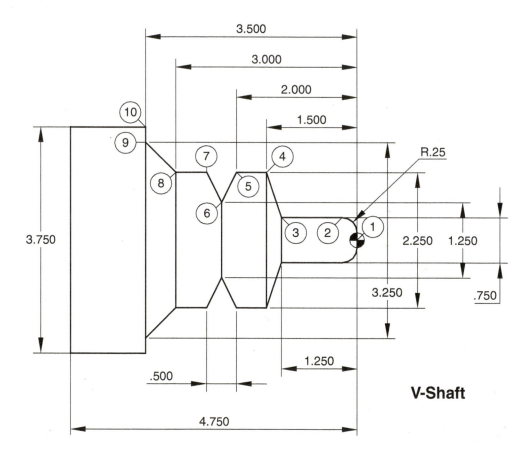

V-Shaft

4. Calculate the X and Z coordinate values of the specified locations on the V-shaft using diameter programming and the absolute system. The X values are diameter values. Place the values in the table.

Position	X	Z	Position	X	Z
1			6		
2			7		
3			8		
4			9		
5					

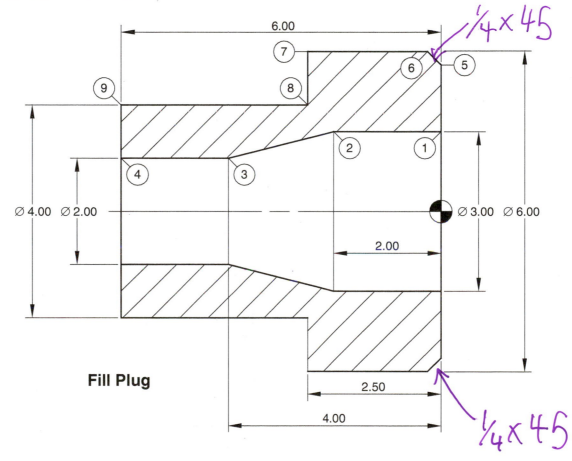

Fill Plug

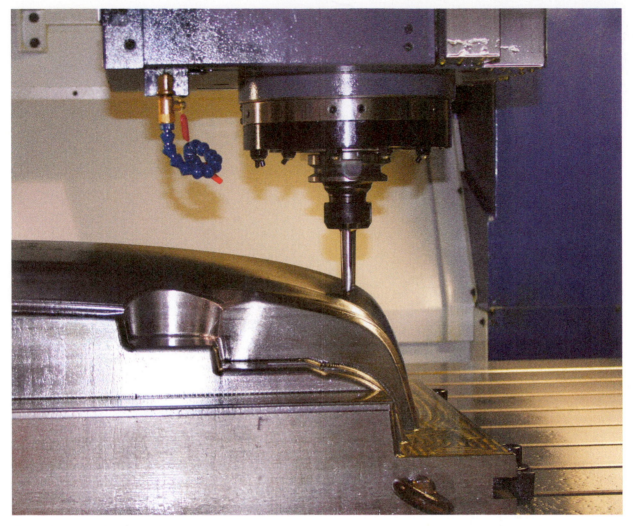

The only way to accurately machine this injection mold insert is to precicely position the cutting tool using the coordinate system.

N10G20G99G40
N20G96S800M3
N30G50S4000
N40T0100M8
N50G00X3.35Z1.25T0101
N60G01X3.25F.002
N70G04X0.5
N80X3.35F.05
N90G00X5.0Z0T0101
01111
N10G20G99G40
N20G96S800M3
N30G50S4000
N40T0100M8
N50G00X3.35Z1.25T0101
N60G01X3.25F.002
N70G04X0.5
N80X3.35F.05

Chapter 3
CNC Math

Objectives

Information in this chapter will enable you to:

- Identify various geometric shapes.
- Apply various geometric principles to solve problems.
- Solve right triangle unknowns.
- Apply trigonometric principles to determine coordinate values.

Technical Terms

acute angle
adjacent angles
angle
arc
bisect
chord
circle
circumference
complementary angles
congruent
cosecant
cosine
cotangent
diagonal
diameter
equilateral triangle

function
hypotenuse
isosceles triangle
obtuse angle
parallel
parallelogram
perpendicular
polygons
proposition
Pythagorean theorem
quadrilateral
radius
rectangle
reflex angle
right angle

right triangle
scalene triangle
secant
segment
sine
square
straight angle
supplementary angles
tangent
transversal
triangle
trigonometric
    functions
trigonometry
vertex

Geometric Terms

The following is a list of geometric terms and their definitions. These terms will be used throughout this chapter and the remainder of this textbook. Study them before continuing.

- *Bisect.* To divide into two equal parts.
- *Congruent.* Having the same size and shape.
- *Diagonal.* Running from one corner of a four-sided figure to the opposite corner.
- *Parallel.* Lying in the same direction but always the same distance apart.
- *Perpendicular.* At a right angle to a line or surface.
- *Segment.* That part of a straight line included between two points.
- *Tangent.* A line contacting a circle at one point.
- *Transversal.* A line that intersects two or more lines.

Angles

An *angle* (∠) is the figure formed by the meeting of two lines at the same point or origin called the *vertex*. See **Figure 3-1**. Angles are measured in degrees (°), minutes ('), and seconds ("). A degree is equal to 1/360 of a circle, a minute is equal to 1/60 of 1°, and a second is equal to 1/60 of 1'.

There are many types of angles, **Figure 3-2**. An *acute angle* is greater than 0° and less than 90°. An *obtuse angle* is greater than 90° and less than 180°. A *right angle* is exactly 90°. A *straight angle* is exactly 180°, or a straight line. A *reflex angle* is greater than 180° and less than 360°.

An angle can also be described by its relationship to another angle. See **Figure 3-3**. *Adjacent angles* are two angles that use a common side. *Complementary angles* are two angles that equal 90°. *Supplementary angles* are two angles that equal 180°, or a straight line.

Polygons

Polygons are figures with many sides that are formed by line segments. Polygons are named according to the number of sides and angles they have. For example, a decagon is a polygon with ten sides; *deca* comes from the Latin word for ten.

Triangles

A *triangle* is a three-sided polygon. There are a number of types of triangles, **Figure 3-4**. A *right triangle* has a 90° (right) angle. An *isosceles triangle* has two equal sides and two equal angles. An *equilateral triangle* has three equal sides; all angles are equal (60°). A *scalene triangle* has three unequal sides and unequal angles.

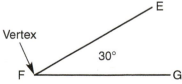

Figure 3-1. Angle defined as EFG.

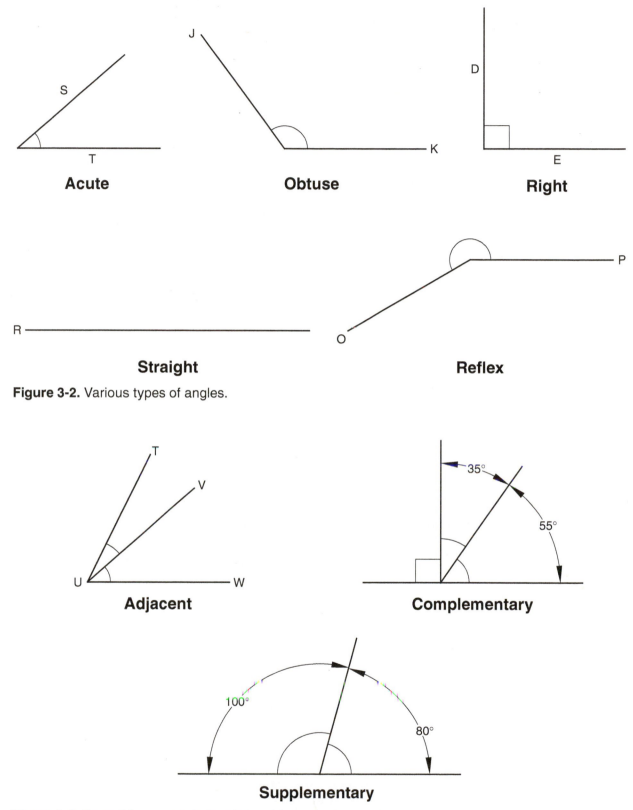

Figure 3-2. Various types of angles.

Figure 3-3. Describing an angle in relationship to another angle.

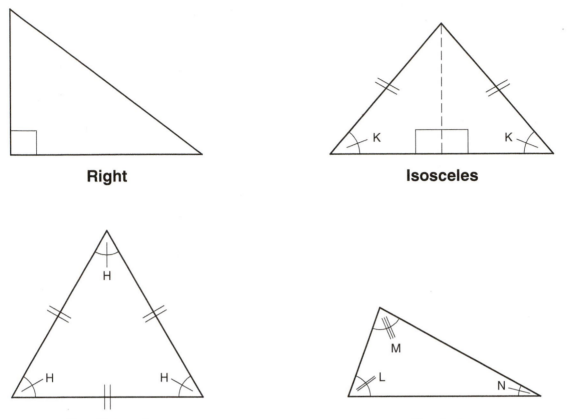

Figure 3-4. Examples of various types of triangles. The sum of the angles in a triangle always equals 180°.

Quadrilaterals

A *quadrilateral* is a polygon with four sides, **Figure 3-5**. A line drawn from one angle (intersecting corner) of a quadrilateral to the opposite angle is called a *diagonal*.

Square

A *square* has four equal sides and four right (90°) angles. The opposite sides of a square are parallel to each other. The diagonals of a square bisect the four angles and each other. The diagonals are equal and perpendicular to each other. The diagonals form congruent angles (equal in size and shape).

Rectangle

A *rectangle* is a quadrilateral with equal opposite sides and four right angles. The opposite sides are parallel to each other. The diagonals are equal, bisect each other, and create two pairs of congruent triangles.

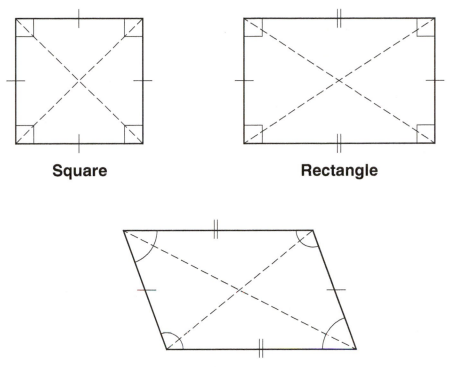

Square

Rectangle

Parallelogram

Figure 3-5. The three types of quadrilaterals. Quadrilaterals have four interior angles that total 360°.

Parallelogram

A *parallelogram* is a quadrilateral with equal opposite sides and equal opposite angles. The diagonals bisect each other and create two pairs of congruent triangles.

Circles

A *circle* is a set of points, located on a plane, that are equidistant from a common central point (center point). See **Figure 3-6**. There are a number of terms that are used to describe various aspects of a circle, **Figure 3-7**.

The *diameter* is the segment that connects two points on a circle and intersects through the center of the circle. The *size* of a circle is its diameter.

The *radius* is a segment that joins the circle center to a point on the circle circumference. Radius has half the value of diameter.

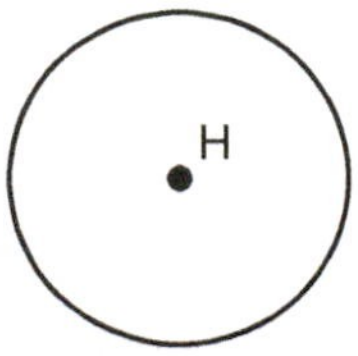

Figure 3-6. All points defining a circle are equidistant from the center point.

H

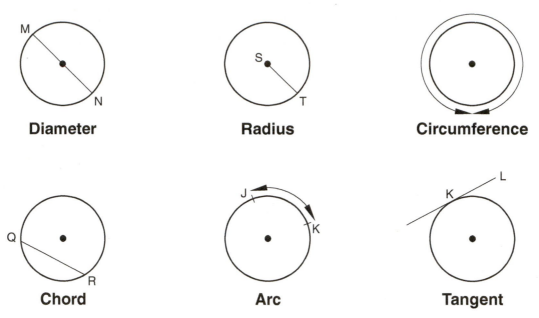

Figure 3-7. Illustrations of various terms relating to a circle.

The *circumference* is the distance around a circle.

A *chord* is a segment that joins any two points on the circumference of a circle.

An *arc* is a curved portion of a circle.

A *tangent* to a circle is a line that intersects a circle at a single point. For example, as shown in Figure 3-7, Line L is tangent to the circle and intersects the circle at Point K.

Propositions

A *proposition* is a statement to be proved, explained, or discussed. Following are a number of geometric propositions.

- **Opposite angles are equal.** When two lines intersect, they form equal angles. Thus, in **Figure 3-8**, Angle 1 equals Angle 3, and Angle 2 equals Angle 4.

- **Two angles are equal if they have parallel corresponding sides.** Thus, in **Figure 3-9**, Angle 1 equals Angle 2.

- **A line perpendicular to one of two parallel lines is perpendicular to the other line.** Thus, in **Figure 3-10**, Lines R and S are perpendicular to Line T.

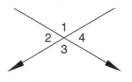

Figure 3-8. Two intersecting lines form four angles with the opposite angles being equal.

- **Alternate interior angles are equal.** If two parallel lines are intersected by a third line (transversal), then alternate interior angles are equal to each other. Thus, in **Figure 3-11**, Angle 3 equals Angle 6, and Angle 4 equals Angle 5.

- **Alternate exterior angles are equal.** When two parallel lines are intersected by a third line (transversal), then alternate exterior angles are equal to each other. Thus, in **Figure 3-11**, Angle 1 equals Angle 8, and Angle 2 equals Angle 7.

- **Corresponding angles are equal.** When two parallel lines are intersected by a third line (transversal), then all corresponding angles are equal. Thus, in **Figure 3-11**, Angle 1 equals Angle 5, Angle 3 equals Angle 7, Angle 2 equals Angle 6, and Angle 4 equals Angle 8.

- **The sum of the interior angles of a triangle is 180°.** Thus, in **Figure 3-12**, Angle 1 plus Angle 2 plus Angle 3 equals 180°.

- **The exterior angle of a triangle is equal to the sum of the two nonadjacent interior angles.** Thus, in **Figure 3-13**, Angle 4 equals Angle 1 plus Angle 2.

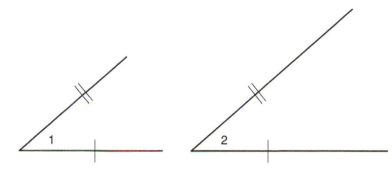

Figure 3-9. Two angles with corresponding parallel sides are equal.

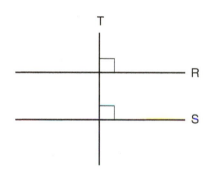

Figure 3-10. A transversal line perpendicular to one parallel line is perpendicular to the other parallel line. Lines R and S are parallel.

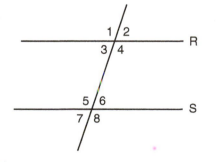

Figure 3-11. Two parallel lines intersected by a third line form alternate angles that are equal to each other. Interior Angles 4 and 5 are equal along with 3 and 6. Exterior Angles 1 and 8 are equal along with 2 and 7.

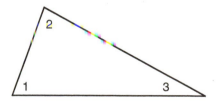

Figure 3-12. Angles 1, 2, and 3 form a triangle totaling 180°.

- **Two angles having their corresponding sides perpendicular are equal.** Thus, in **Figure 3-14**, Angle 1 equals Angle 2.

- **A line taken from a point of tangency to the center of a circle is perpendicular to the tangent.** Thus, in **Figure 3-15**, Line TW is perpendicular to Line UV.

- **Two tangent lines drawn to a circle from the same exterior point cause the corresponding segments to be equal in length.** Thus, in **Figure 3-16**, Segment ML equals Segment MN.

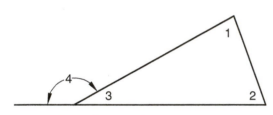

Figure 3-13. Angle 4, an exterior angle, equals the sum of Angles 1 and 2, which are nonadjacent interior angles.

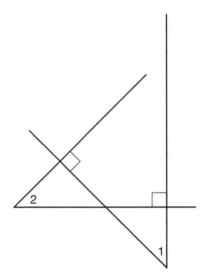

Figure 3-14. Angle 1 is equal to Angle 2 because their corresponding sides are perpendicular to each other.

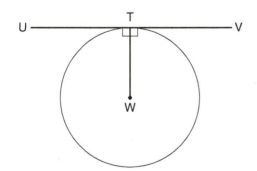

Figure 3-15. Line TW, developed from the tangency point to the circle's center, is perpendicular to tangent Line UV.

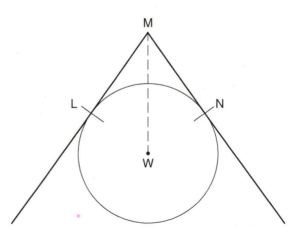

Figure 3-16. Lines MN and ML are tangent to the circle and share an endpoint and are, therefore, equal in length.

Trigonometry

Trigonometry is the area of mathematics that deals with the relationship between the sides and angles of a triangle. Triangles are measured to find the length of a side (leg) or to find the number of degrees in an angle. In CNC machining, trigonometry is used to determine tool location relative to part geometry.

Trigonometry deals with the solution of triangles, primarily the right triangle. See **Figure 3-17**. A right triangle has one angle that is 90° (Angle c), and the sum of all angles equals 180°. Angles a and b are acute angles, which means they each are less than 90°. Angles a and b are complementary angles, which means they total 90° when added.

The three sides of a triangle are called the hypotenuse, side opposite, and side adjacent. Side C is called the *hypotenuse*, because it is opposite the right angle. It always is the longest side.

Sides A and B are either opposite to or adjacent to either of the acute angles. It depends on which acute angle is being considered. Side A is the side opposite Angle a, but is the side adjacent to Angle b. Side B is the side opposite Angle b, but is the side adjacent to Angle a. For example, when referring to Angle b, Side A is adjacent and Side B is opposite. Or, when referring to Angle a, Side B is adjacent and Side A is opposite.

As stated earlier in this chapter, angles are usually measured in degrees, minutes, and seconds, **Figure 3-18**. There are 360° in a circle, 60′ in a degree, and 60″ in a minute. As an example, 31 degrees, 16 minutes, and 42 seconds is written as 31°16′42″. Angles can also be given in decimal degrees, such as 34.1618 (34°9′42″).

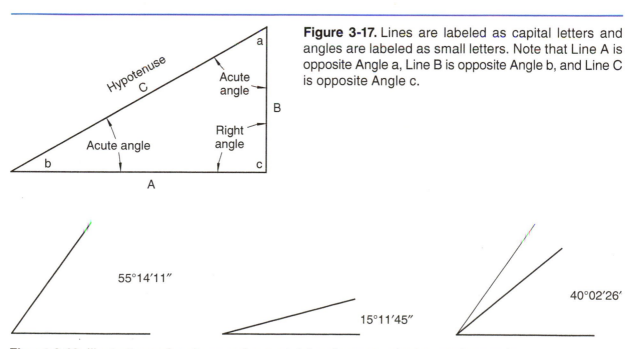

Figure 3-17. Lines are labeled as capital letters and angles are labeled as small letters. Note that Line A is opposite Angle a, Line B is opposite Angle b, and Line C is opposite Angle c.

Figure 3-18. Illustrations of various angles containing degrees, minutes, and seconds.

Angles can be added by aligning the degrees, minutes, and seconds and adding each column separately. When totals for the minutes or seconds columns add up to 60 or more, subtract 60 (or 120, if appropriate) from that column, then add 1 (or 2, if appropriate) to the next column to the left (the higher column).

16°	33′	14″
5°	17′	16″
38°	55′	49″
59°	105′	79″
	-60	**-60**
	45′	19″
+1	**+1**	
60°	46′	19″

In the example, the total is 59°105′79″ when the angles are added. Since 79″ equals 1′19″ and 105′ equals 1°45′, the final answer is 60°46′19″.

When subtracting angles, place the degrees, minutes, and seconds under each other and subtract the separate columns. If not enough minutes or seconds exist in the upper number of a column, then borrow 60 from the next column to the left of it and add it to the insufficient number.

55°	14′	11″	**borrow 60″→**	55°	13′	71″
−15°	11′	45″		−15°	11′	45″
				40°	2′	26″

Since 11″ is smaller than 45″, 60″ must be borrowed from 14′. When 15°11′45″ is subtracted from 55°13′71″, the final answer is 40°2′26″.

Using Trigonometry

Trigonometry is the most valuable mathematical tool used by a programmer for calculating cutter or tool nose locations. Trigonometric functions are absolute values derived from the relationships existing between angles and sides of a right triangle. A *function* is a magnitude (size or dimension) that depends upon another magnitude. For example, a circle's circumference is a function of its radius, since the circle size depends on the extent of its radius value.

In the triangle shown in **Figure 3-19**, A/B is the ratio of two sides and therefore a function of Angle d. As Angle d increases to the dashed line, the function will change from A/B to E/B. This shows that the ratio of two sides of a triangle depends on the size of the angles of the triangle.

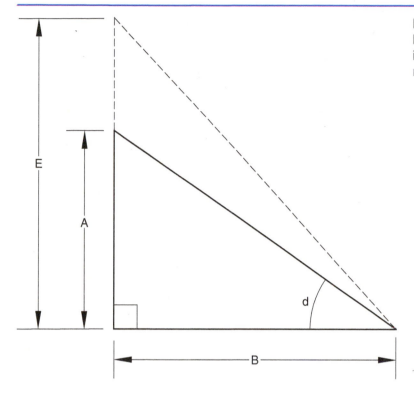

Figure 3-19. As Line A increases in length to become Line E, Angle d increases in value, when Line B remains the same.

Trigonometric Functions

Since there are three sides (legs) to a triangle, there exist six different ratios of sides. These ratios are the six *trigonometric functions* of sine, cosine, tangent, cotangent, secant, and cosecant. Each ratio is named from its relationship to one of the acute angle in a right triangle. The right angle is never used in calculating functions. A function is obtained by dividing the length of one side by the length of one of the other sides. These functions can be found in math texts and many references, such as *Machinery's Handbook*. Special books also exist that give primarily trigonometric tables and values. In addition, many calculators can compute trigonometric values. **Figure 3-20** is a partial table of trigonometric functions covering 33°. To find the cosine of 33°58′, read down the minute column to 58 minutes, then read across the row. Under the column labeled cosine, you find value 0.82936, which is cosine 33°58′.

Trigonometric Functions for Angles

Angle	Sine	Cosine	Tangent	Cotangent	Secant	Cosecant
33°0′	0.54464	0.83867	0.64941	1.53986	1.19236	1.83608
33°1′	0.54488	0.83851	0.64982	1.53888	1.19259	1.83526
33°2′	0.54513	0.83835	0.65024	1.53791	1.19281	1.83444
33°3′	0.54537	0.83819	0.65065	1.53693	1.19304	1.83362
33°4′	0.54561	0.83804	0.65106	1.53595	1.19327	1.83280
33°5′	0.54586	0.83788	0.65148	1.53497	1.19349	1.83198
33°6′	0.54610	0.83772	0.65189	1.53400	1.19372	1.83116
33°7′	0.54635	0.83756	0.65231	1.53302	1.19394	1.83034
33°8′	0.54659	0.83740	0.65272	1.53205	1.19417	1.82953
33°9′	0.54683	0.83724	0.65314	1.53107	1.19440	1.82871
33°10′	0.54708	0.83708	0.65355	1.53010	1.19463	1.82790
33°11′	0.54732	0.83692	0.65397	1.52913	1.19485	1.82709
33°12′	0.54756	0.83676	0.65438	1.52816	1.19508	1.82627
33°49′	0.55654	0.83082	0.66986	1.49284	1.20363	1.79682
33°50′	0.55678	0.83066	0.67028	1.49190	1.20386	1.79604
33°51′	0.55702	0.83050	0.67071	1.49097	1.20410	1.79527
33°52′	0.55726	0.83034	0.67113	1.49003	1.20433	1.79449
33°53′	0.55750	0.83017	0.67155	1.48909	1.20457	1.79371
33°54′	0.55775	0.83001	0.67197	1.48816	1.20480	1.79293
33°55′	0.55799	0.82985	0.67239	1.48722	1.20504	1.79216
33°56′	0.55823	0.82969	0.67282	1.48629	1.20527	1.79138
33°57′	0.55847	0.82953	0.67324	1.48536	1.20551	1.79061
33°58′	0.55871	0.82936	0.67366	1.48442	1.20575	1.78984
33°59′	0.55895	0.82920	0.67409	1.48349	1.20598	1.78906

Figure 3-20. Partial table showing values of the six trigonometric functions sine, cosine, tangent, cotangent, secant, and cosecant as they relate to 33° and various minutes.

The six trigonometric functions are defined relative to the relationships between two sides of the right triangle, **Figure 3-21**. These relationships are:

- *Sine* **(sin).** The ratio of the opposite side to the hypotenuse.

$$\text{Sine } a = \frac{\text{Opposite side}}{\text{Hypotenuse}} = \frac{A}{C}$$

- *Cosine* **(Cos).** The ratio of the adjacent side to the hypotenuse.

$$\text{Cosine } a = \frac{\text{Adjacent side}}{\text{Hypotenuse}} = \frac{B}{C}$$

- *Tangent* **(Tan).** The ratio of the opposite side to the adjacent side.

$$\text{Tangent } a = \frac{\text{Opposite side}}{\text{Adjacent side}} = \frac{A}{B}$$

- *Cotangent* **(Cot).** The ratio of the adjacent side to the opposite side. It is the reciprocal of the tangent function.

$$\text{Cotangent } a = \frac{\text{Adjacent side}}{\text{Opposite side}} = \frac{B}{A}$$

- *Secant* **(Sec).** The ratio of the hypotenuse to the adjacent side. It is the reciprocal of the cosine function.

$$\text{Secant } a = \frac{\text{Hypotenuse}}{\text{Adjacent side}} = \frac{C}{B}$$

- *Cosecant* **(Csc).** The ratio of the hypotenuse to the opposite side. It is the reciprocal of the sine function.

$$\text{Cosecant } a = \frac{\text{Hypotenuse}}{\text{Opposite side}} = \frac{C}{A}$$

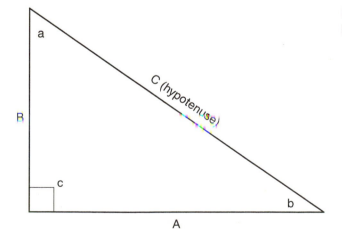

Figure 3-21. The hypotenuse is always the longest side (leg) of a right triangle.

The six functions given are related to Angle a, but also can be applied to Angle b as well. Therefore, Sin b = B/C, Cos b = A/C, etc., shows that any function of Angle a is equal to the cofunction of Angle b. From that relationship, the following are derived:

- sin a = A/C = cos b
- cos a = B/C = sin b
- tan a = A/B = cot b
- cot a = B/A = tan b
- sec a = C/B = csc b
- csc a = C/A = sec b

With Angle a and Angle b being complementary, the function of any angle is equal to the cofunction of its complementary angle. Therefore, sin 70° = cos 20°, and tan 60° = cot 30°.

Working with Triangles

In programming, an individual will be working with various applications of radii, such as cutter radius, arc radius, circle radius, and corner radius. At times, the radius the programmer works with will appear like the triangle in **Figure 3-22**, where the long leg of the triangle is the radius. At other times, the triangle will appear like the triangle in **Figure 3-23**, where a leg will be the radius. When dealing with triangles, a programmer must recognize the configuration being worked with. There

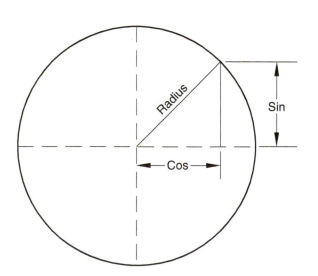

Figure 3-22. Programmers will sometimes use the radius value as the hypotenuse when determining location values.

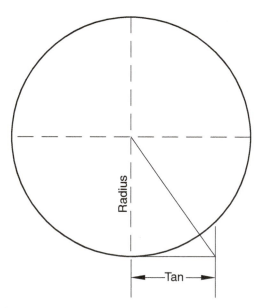

Figure 3-23. Programmers may use the radius value as a leg to solve the other missing values of the right triangle.

are a number of applications where a programmer will use the radius and triangle to determine distances. In **Figure 3-24**, the radius and triangle are applied to determine a bolt circle, tool path, and intersection. In **Figure 3-25**, the radius and triangle are applied to determine a cutter path.

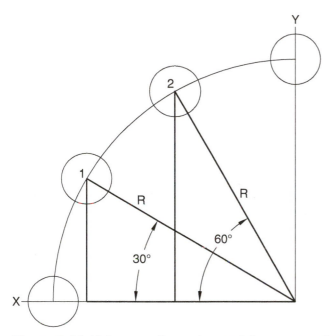

Position 1	Position 2
R = 1.5″	R = 1.5″
Angle from the horizontal axis = 30°	Angle from the horizontal axis = 60°
Y = sin 30 (1.5″)	Y = sin 60 (1.5″)
Y = 0.5 (1.5″)	Y = 0.866 (1.5″)
Y = 0.75″	Y = 1.299″
X = cos 30 (1.5″)	X = cos 60 (1.5″)
X = 0.866 (1.5″)	X = 0.5 (1.5″)
X = 1.299″	X = 0.75″

Figure 3-24. Using a radius value and the construction of a right triangle to determine hole locations. Note: Angle values are used to solve remaining leg values.

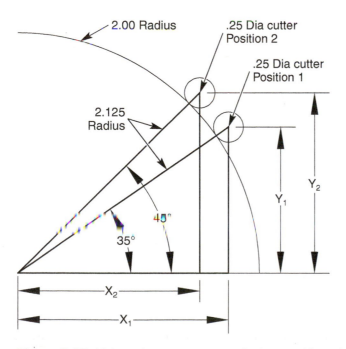

Position 1	Position 2
R = 2.0″	R = 2.0″
Cutter diameter = 0.250″	Cutter diameter = 0.250″
Angle from the horizontal axis = 35″	Angle from the horizontal axis = 45″
Y_1 = sin 35 (2.125″)	Y_2 = sin 45 (2.125″)
Y_1 = 0.5735 (2.125″)	Y_2 = 0.707 (2.125″)
Y_1 = 1.2186″	Y_2 = 1.502″
X_1 = cos 35 (2.125″)	X_2 = cos 45 (2.125″)
X_1 = 0.819 (2.125″)	X_2 = 0.707 (2.125″)
X_1 = 1.740″	X_2 = 1.502″

Figure 3-25. Using trigonometry to calculate tool locations.

There are a number of situations where triangles can be applied to determine a cutter path. To plot a cutter path, the cutter radius is added or subtracted from the part outline. The cutter path is the path in which the centerline of the spindle moves along the plane, staying away from the part by the amount of the tool radius. To cut a 90° corner, the cutter moves past the edges of the part a distance equal to the cutter radius. See **Figure 3-26**. To cut an acute (less than 90°) angle, the cutter moves past the corner of the workpiece equal to the distances represented by the X dimension of the shaded triangle in **Figure 3-27**. To cut an obtuse (greater than 90°) angle, as shown in **Figure 3-28**, the same formula is used. Notice that the distance the cutter has to travel beyond the end of the part is greater than the cutter radius for acute angles and less than the cutter radius for obtuse angles.

Figure 3-26. The location of a cutter when it is about to make a 90° move is shown. The radius of the cutter is used when determining the cutter location from the corner of the workpiece.

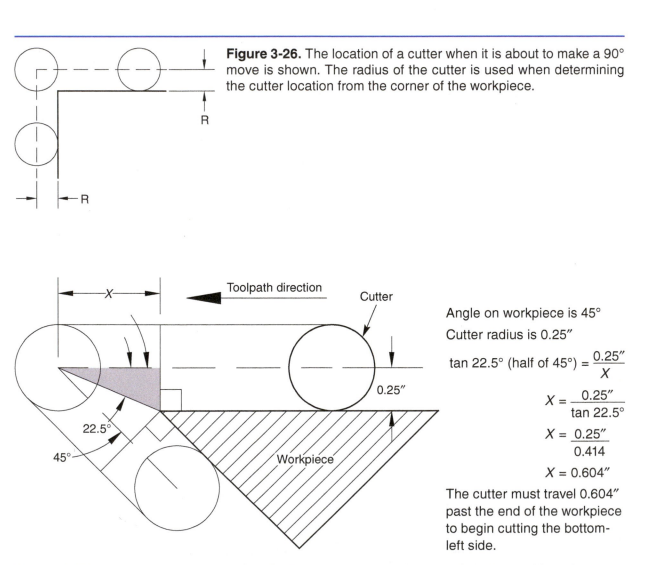

Angle on workpiece is 45°

Cutter radius is 0.25″

$$\tan 22.5° \text{ (half of 45°)} = \frac{0.25″}{X}$$

$$X = \frac{0.25″}{\tan 22.5°}$$

$$X = \frac{0.25″}{0.414}$$

$$X = 0.604″$$

The cutter must travel 0.604″ past the end of the workpiece to begin cutting the bottom-left side.

Figure 3-27. Illustration showing the triangle that must be solved to calculate the position of a cutter when cutting an acute angle on a workpiece.

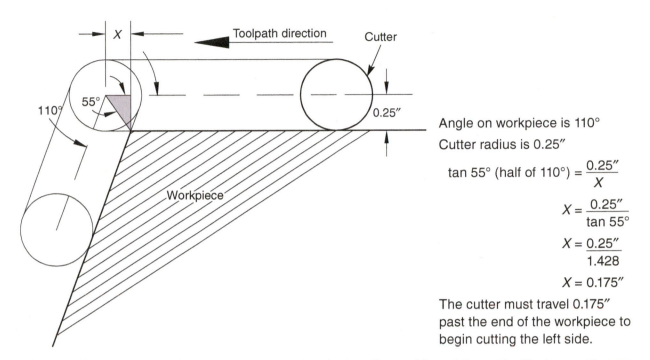

Angle on workpiece is 110°

Cutter radius is 0.25″

$$\tan 55° \text{ (half of 110°)} = \frac{0.25″}{X}$$

$$X = \frac{0.25″}{\tan 55°}$$

$$X = \frac{0.25″}{1.428}$$

$$X = 0.175″$$

The cutter must travel 0.175″ past the end of the workpiece to begin cutting the left side.

Figure 3-28. The triangle that must be solved to calculate the position of the end mill when cutting an obtuse angle on a workpiece.

Pythagorean Theorem

The *Pythagorean theorem* states a special relationship that exists among the three sides of a right triangle. It states that *the length of the hypotenuse squared equals the sum of the squares of the other two side lengths.* So, if the lengths of any two sides of a right triangle are given, the length of the third side can be calculated by using the Pythagorean theorem:

$$A^2 + B^2 = C^2$$

In **Figure 3-29**, Side C is equal to 5 and Side B is equal to 3. The value for A (the third side of the triangle) can be determined by using the formula $C^2 = A^2 + B^2$. To solve for A, substitute the known values into the formula to get $5^2 = A^2 + 3^2$, then square the values to get $25 = A^2 + 9$. Next, isolate the unknown variable by subtracting 9 from both sides of the equation to get $16 = A^2$. Finally, take the square root of both sides of the equation, to get $4 = A$. So, the length of Side A is 4.

To cut a 90° rounded corner on a workpiece, we can use the Pythagorean theorem to plot the toolpath of the cutter. See **Figure 3-30**. The radius on the workpiece is 1″. The cutter diameter is 0.25″ (0.125″ radius).

To cut partial arcs, we can use a combination of trigonometric functions and the Pythagorean theorem to plot the positions of the cutter. See **Figure 3-31**.

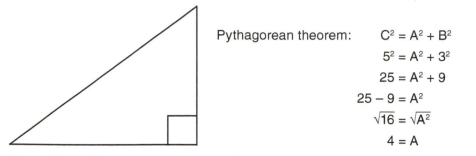

Pythagorean theorem: $C^2 = A^2 + B^2$

$5^2 = A^2 + 3^2$

$25 = A^2 + 9$

$25 - 9 = A^2$

$\sqrt{16} = \sqrt{A^2}$

$4 = A$

Figure 3-29. Cutter locations when cutting a 90° radius corner on a workpiece.

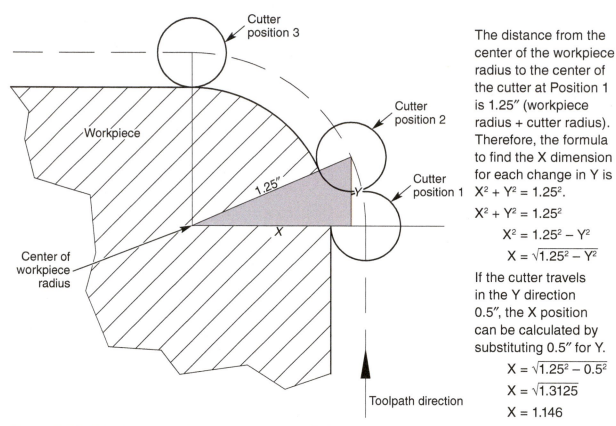

The distance from the center of the workpiece radius to the center of the cutter at Position 1 is 1.25″ (workpiece radius + cutter radius). Therefore, the formula to find the X dimension for each change in Y is $X^2 + Y^2 = 1.25^2$.

$$X^2 + Y^2 = 1.25^2$$
$$X^2 = 1.25^2 - Y^2$$
$$X = \sqrt{1.25^2 - Y^2}$$

If the cutter travels in the Y direction 0.5″, the X position can be calculated by substituting 0.5″ for Y.

$$X = \sqrt{1.25^2 - 0.5^2}$$
$$X = \sqrt{1.3125}$$
$$X = 1.146$$

Figure 3-30. This illustration shows how to use the Pythagorean theorem to calculate the cutter position as it creates a corner radius.

To start, we need to calculate the position of the cutter in Position 2 based on the center of the workpiece radius. We know that Y = 1.00″ and H = 2.25. Therefore, X can be calculated using Pythagorean's theorem.

$$X^2 + 1^2 = 2.25^2$$
$$X^2 = 2.25^2 - 1^2$$
$$X = \sqrt{2.25^2 - 1^2}$$
$$X = \sqrt{5.0625 - 1}$$
$$X = 2.0156$$

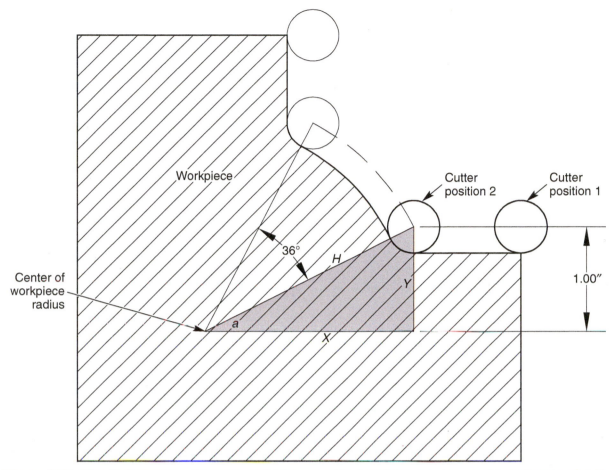

Figure 3-31. The Pythagorean theorem and trigonometric functions can be used together to plot the toolpath of a cutter.

We can use trigonometric functions to plot the positions of the cutter as it travels along the 2.00″ radius in 1° increments. To do this we have to find angle *a*.

$$\sin a = \frac{1}{2.25″}$$

$$a = \sin^{-1}(.4444)$$

$$a = 26°$$

Now we can calculate distance X for each degree that angle *a* increases until it reaches 62° (26° + 36°).

$$\sin 27 = \frac{Y}{2.25″}$$

$$\sin 27 \, (2.25) = Y$$

$$1.021 = Y$$

$$\sin 28 \, (2.25) = Y$$

$$1.056 = Y$$

Summary

Various geometric principles relating to triangles, quadrilaterals, and circles are important to learn. Many of these principles are applied to obtain needed data for calculating tool locations. There are several propositions relating to angles that should be learned by an individual applying math to calculate tool location.

Using trigonometry to solve for the various parts of a triangle is an important concept. Tool location is often determined by using the trigonometric functions. There are six trigonometric functions and each is defined relative to the relationships between two sides of the right triangle.

Chapter Review

Answer the following questions. Write your answers on a separate sheet of paper.

1. List the complementary angles for the following.
   a. 62°
   b. 41°
   c. 14°32′
2. List the supplementary angles for the following.
   a. 76°
   b. 167°
   c. 145°25′15″

3. Determine the values of the angles shown in the figure below. State the propositions used in the problem. Lines 1 and 2 are parallel, and Lines 3 and 4 are parallel.

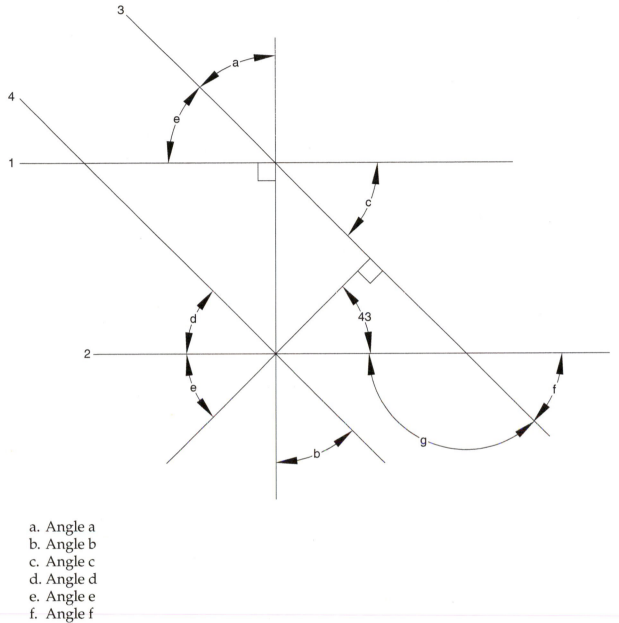

   a. Angle a
   b. Angle b
   c. Angle c
   d. Angle d
   e. Angle e
   f. Angle f
   g. Angle g

4. Using a table or calculator, determine the value of each function listed.
   a. Tan 33.15°
   b. Sin 23°
   c. Cos 26.6°
   d. Cot 41°

5. Using the triangle below, solve for R and S.

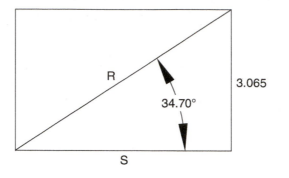

6. Using the triangle below, solve for U and t.

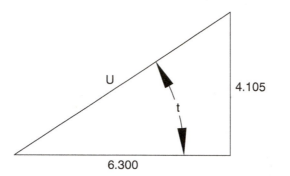

7. Using the triangle below, solve for D and e.

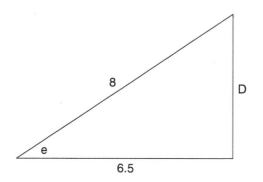

Activities

1. Using the print, calculate the value of the positions identified by Points 1–11. Point 6, Point 10, and the center point of the 1.25" radius arc are collinear. It may be necessary to use the formulation of right triangles and trigonometry to calculate certain positions. Place the values determined into a table similar to the one shown.

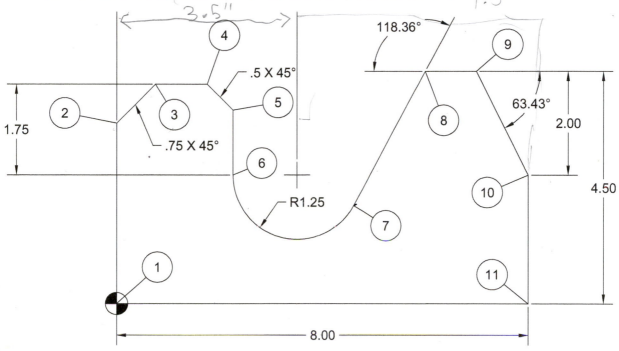

Location	X	Y
1		
2		
3		
4		
5		
6		
7		
8		
9		
10		
11		

Right Triangle Formulas and Calculator Steps

To find...	...and you know...	...perform these calculator steps.	Formulas
angle B	sides a & b	b ÷ a 2nd SIN⁻¹	$b/a = \text{Sin } B$
angle B	sides a & c	c ÷ a 2nd COS⁻¹	$c/a = \text{Cos } B$
angle B	sides b & c	b ÷ c 2nd TAN⁻¹	$b/c = \text{Tan } B$
angle C	sides a & b	b ÷ a 2nd COS⁻¹	$b/a = \text{Cos } C$
angle C	sides a & c	c ÷ a 2nd SIN⁻¹	$c/a = \text{Sin } C$
angle C	sides b & c	c ÷ b 2nd TAN⁻¹	$c/b = \text{Tan } C$
side a	sides b & c	b x^2 + c x^2 2nd √	$\sqrt{b^2 + c^2}$
side a	side c & angle C	c ÷ (C SIN) ENTER	$\dfrac{c}{\sin C}$
side a	side c & angle B	c ÷ (B COS) ENTER	$\dfrac{c}{\cos B}$
side a	side b & angle B	b ÷ (B SIN) ENTER	$\dfrac{b}{\sin B}$
side a	side b & angle C	b ÷ (C COS) ENTER	$\dfrac{b}{\cos C}$
side b	sides a & c	a x^2 + c x^2 2nd √	$\sqrt{a^2 + c^2}$
side b	side a & angle B	B SIN × a	$a \times \text{Sin } B$
side b	side a & angle C	C COS × a	$a \times \text{Cos } C$
side b	side c & angle B	B TAN × c	$c \times \text{Tan } B$
side b	side c & angle C	c ÷ (C TAN) ENTER	$\dfrac{c}{\tan C}$
side c	sides a & b	a x^2 + b x^2 2nd √	$\sqrt{a^2 + b^2}$
side c	side a & angle B	B COS × a	$a \times \text{Cos } B$
side c	side a & angle C	C SIN × a	$a \times \text{Sin } C$
side c	side b & angle B	b ÷ (B TAN) ENTER	$\dfrac{b}{\tan B}$
side c	side b & angle C	C TAN × b	$b \times \text{Tan } C$

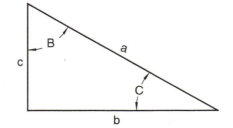

N10G20G99G40
N20G96S800M3
N30G50S4000
N40T0100M8
N50G00X3.35Z1.25T0101
N60G01X3.25F.002
N70G04X0.5
N80X3.35F.05
N90G00X5.0Z0T0101
O1111
N10G20G99G40
N20G96S800M3
N30G50S4000
N40T0100M8
N50G00X3.35Z1.25T0101
N60G01X3.25F.002
N70G04X0.5
N80X3.35F.05

Chapter 4
Machining Centers

Objectives

Information in this chapter will enable you to:

- Identify the axes on vertical and horizontal machining centers.
- List various work setup items used on machining centers.
- Name three types of automatic tool-changing systems.
- Distinguish between rotary and indexing devices.
- State the purpose of tool length offsets.
- Understand the difference between plus and minus direction trim.
- State several reasons for using tool radius offsets.

Technical Terms

angle plates	indexers	S-word
carousel	machining center	straps
clamps	manual data input	T-bolts
coolant	(MDI)	tool length offsets
drawbar	modular fixtures	tool radius offsets
dwell	parallels	turret
F-command	rotary tables	V-blocks
fixtures	step blocks	vertical machining
horizontal machining	subplates	center
center	support jacks	

Machining Centers

Most machine shops with CNC equipment have at least one machining center. A *machining center* is a milling machine that uses programmed commands, has computer controlled movement in three or more axes, and is able to perform automatic tool changes. Machining centers perform many forms of hole-making operations, in addition to all types of milling procedures. The principal concept behind these machine tools is that the spindle rotates and the workpiece remains stationary. There are two broad categories of machining centers: vertical and horizontal.

Vertical Machining Centers

On a *vertical machining center,* the machine spindle is in a vertical position, perpendicular to the machine table, **Figure 4-1**. This machine is a saddle type construction with sliding bedways and a sliding vertical headstock. Vertical machining centers are similar to vertical milling machines, but use computer numerical control positioning and automatic tool changers to produce complex machine parts, in usually a single setup. Drilling, tapping, boring, end milling, and face milling are the typical machining operations performed on the vertical machining center.

Vertical machining centers can vary in design, depending on operator involvement. A numerical control (NC) vertical milling machine may require tools to be loaded manually and speeds to be selected manually, whereas the CNC (computer numerical control) machining centers control all functions—such as speeds, feeds, coolant, and tool loading—through computer programmed commands.

The vertical machining center is probably the easiest machining center to operate, because it is very similar in construction to the manual vertical mill that most seasoned mill operators have used. It machines one side of the part at a time, unless it uses a rotary holding device.

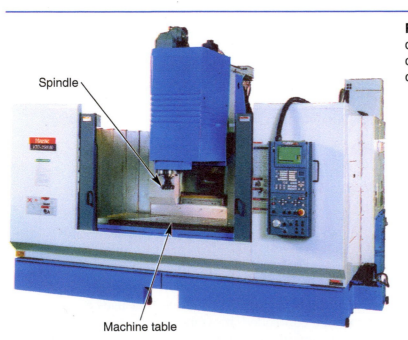

Spindle

Machine table

Figure 4-1. A vertical machining center. This is a vertical traveling column machine with a 24 tool capacity. (Mazak)

Directions of movement

Vertical machining centers allow a minimum of three directions of axis movement, **Figure 4-2**. The table can move left or right and in or out as viewed by an operator facing the front of the machine. The tool-carrying spindle (also called the *head* or *quill*) moves up or down. Left/right movement constitutes motion in the X axis. In/out movement constitutes motion in the Y axis; up or down movement of the spindle provides Z axis motion.

Movement in five axes allows the machining of complex shaped parts using one or fewer setups as compared to conventional manual machining. See **Figure 4-3**. Because of their ability to perform machining operations on all sides of a workpiece, such machining centers are very popular. The five-axis machining center can perform simultaneous movement in these directions:

- *Y axis.* Linear movement.
- *X axis.* Linear movement.
- *A axis.* Tilt/contour spindle.
- *B axis.* Rotary table.

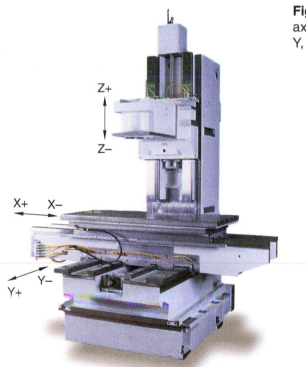

Figure 4-2. Many vertical machining centers have three axes of movement, typically identified with the letters X, Y, and Z. (Bridgeport Machines)

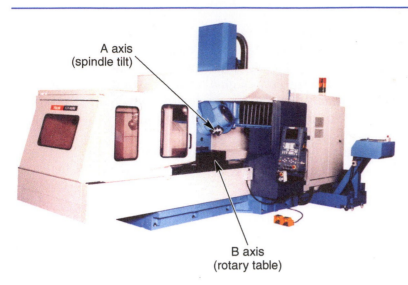

A axis
(spindle tilt)

B axis
(rotary table)

Figure 4-3. A five-axis machining center adds an A axis (tilting of the spindle) and a B axis (rotary workholding table) to the usual X, Y, and Z axes. (Mazak)

Horizontal Machining Centers

The *horizontal machining center* has the machine spindle mounted with its axis in a horizontal position. While the horizontal machining center is much more flexible than the vertical machining center, it is more difficult to operate and set up. The reasons a horizontal machining center is more difficult to operate are:

- Programs are longer and usually more complicated.
- Setups are more sophisticated, often involving special workholding fixtures.
- Tooling requirements are usually more demanding.

Horizontal machining centers perform the same functions as vertical machining centers. With the use of indexing or rotary devices, however, multiple sides of a workpiece, or multiple parts, can be machined on one setup. Tooling, speeds, feeds, coolant, and various other functions are controlled by the computer. There are two types of horizontal machining centers: traveling-column and fixed-column.

The *traveling-column* type is often equipped with two worktables on which the workpieces can be mounted. One workpiece can be machined while the operator loads another table. The *fixed-column* type is fitted with a pallet shuttle. The pallet is a removable table on which the workpiece or workpieces are held. After the machining is completed, the job and pallet are transported to a shuttle that then shifts or rotates, moving a new pallet and workpiece into position for machining. See **Figure 4-4**.

Directions of movement

Horizontal machining centers use the same axes as the vertical machines, but change designations somewhat because of the spindle orientation. Table movement left/right is still the X axis but the Y axis is

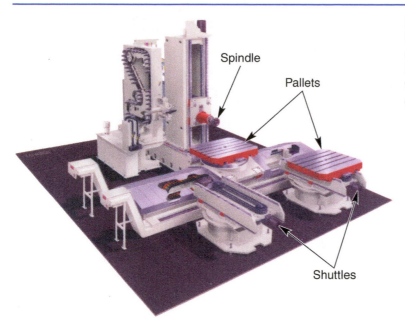

Spindle

Pallets

Shuttles

Figure 4-4. This fixed column horizontal machining center with a double pallet system allows a part to be loaded or unloaded on one pallet while a part is being machined on the other pallet. (Giddings and Lewis)

the movement of the head up/down. Table motion toward/away from the spindle is the Z axis. Frequently an indexer or rotary table provides a fourth axis called the B axis, which is rotary travel about the Y axis.

Workholding Devices

Workholding devices and methods have an important role in the setup and operation of the machining center. Safety and setup rigidity are the most important items to consider when locating and fastening workpieces to be machined. The workpiece must be properly positioned and fastened to the table. Huge cutting forces are exerted in machining operations affected by speeds, feeds, and type of cutting application. Carelessness in setting up and fastening workpieces can lead to broken tooling, insecure clamping, and dislocated workpieces. The unwanted result of such carelessness can be a scrapped part, a broken tool, or the worst outcome, an injury or death.

The operator should check that the holding device and workpiece are free of dirt, chips, and burrs. The holding device and workpiece must be properly located on the table according to program or setup instructions. Before running a part program for machining, make sure the part is secure. Shop setup personnel must have a thorough knowledge of the use of available holding and locating components. Fixture designers, especially, must be aware of all the items available for workholding applications.

Vises

The vise, **Figure 4-5**, is the most common holding device. It has fixed and movable parallel jaws that can hold a workpiece. The plain vise is

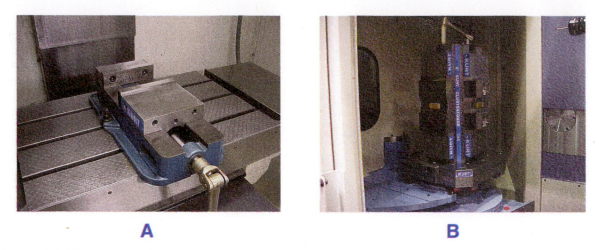

A   **B**

Figure 4-5. Vises. A—A vise mounted on the table of vertical machining center. B—Vises mounted on a tooling block on horizontal machining center. A tooling block allows the mounting of several vises or other fixturing to hold several workpieces.

bolted directly to the machine table or a subplate. Vises are used for holding workpieces with parallel surfaces. Swivel-base vises allow the vise to be rotated 360°. Air or hydraulic vises are used for high-production situations. Multiple vises can be used in a setup; they can be mounted in various positions on tooling towers.

Angle Plates

Angle plates are used to hold irregularly shaped work with a flat surface or hold work at a right angle to the axis of tool travel. See **Figure 4-6**. An angle plate is an L-shaped device with slots and tapped holes that provide a method for fastening parts. Parts or fixtures can be fastened to angle plates to provide more part planes to be accessible to machining.

Subplates

Subplates are flat, ground plates with accurately located tapped and reamed holes. Subplates provide a means for locating and fastening workpieces. Accurate placement on a machine table relative to machine zero provides the programmer with known locating positions.

Multiple parts and fixtures are often mounted on these subplates. Sometimes, subplates are referred to as *locating plates*.

V-blocks

V-blocks, **Figure 4-7**, are generally used in pairs to support cylindrical workpieces. They are fastened down with clamps or straps or could be held in a vise. V-blocks come in various shapes and sizes. Special V-blocks have clamps attached to them that hold down the workpiece. In some cases, both the workpiece and V-block are held down by clamps.

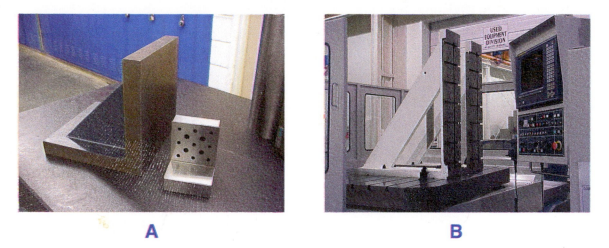

Figure 4-6. Angle plates. A—Small angle plates can be mounted on a table to hold a workpiece. Angle plates can have tapped holes in them to provide easy mounting of workpieces. B—Large angle plates mounted on the table of a horizontal boring and milling machine. Notice the T-slots that are machined into the angle plates to aid in mounting workpieces or fixtures.

Figure 4-7. V-blocks are used to hold cylindrical workpieces.

Direct Table Mounting

Workpieces that are odd shaped or very large are sometimes bolted directly to the machine table. A variety of holding accessories are used to fasten the workpiece.

- *Step blocks.* These blocks serve as support for clamps and for clamp height adjustment. Steps can be flat or manufactured with serrated steps used with special serrated clamps. See **Figure 4-8**.

- *Clamps* or *straps.* These include plain bar clamps, spring-loaded L-clamps, U-clamps, offset clamps, and universal clamps. See **Figure 4-9**.

- *Support jacks.* This holding accessory is adjustable to provide support under a workpiece to prevent distortion caused by clamping. There are many variations of support jacks that are used.

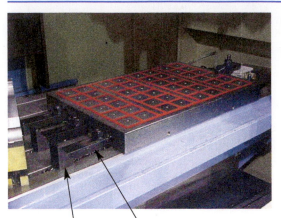

Figure 4-8. Step blocks used with step clamps to hold down a magnetic table.

Step block   Clamp

Standard heel clamp
A

Gooseneck clamp
B

Toe clamp
C

Figure 4-9. Examples of clamps.

- *Parallels.* These precision-ground flat, square, or rectangle steel or cast iron bars are usually used in vises to support and raise a workpiece. See **Figure 4-10**.
- *T-bolts.* These special bolts engage worktable T-slots, and are used with straps and clamps. T-nuts with studs of various lengths are sometimes used instead of T-bolts. See **Figure 4-11**.

Fixtures

Fixtures are specially designed and built holding devices used exclusively to hold and locate a specific part or parts. See **Figure 4-12**. They are typically found in high production situations and used for oddly configured parts that cannot be fastened in a vise. Fixtures are designed to provide quick workpiece loading and unloading.

Figure 4-10. A set of thin parallels. Parallels come in pairs and different designs. Some have holes to reduce weight.

T-bolt

Figure 4-11. T-bolts are used with step clamps to hold down this fixture.

Figure 4-12. A horizontal machine with a vertical fixture holding two parts. (The Goss & DeLeeuw Machine Company)

Modular fixtures

Modular fixtures are specialty components designed for clamping, **Figure 4-13**. They are assembled on a special plate that has accurately located tapped holes to which components are attached. Fixtures can be assembled and disassembled and components can be reused for different setups.

Figure 4-13. A vertical modular fixture being used to hold down a cylindrical aluminum casting.

Special Functions

Special functions refer to the special features found on CNC machines. These functions are within a program that are activated and controlled by the machine controller.

Spindle Speed

Spindle speeds on CNC machining centers are programmed in revolutions per minute (rpm). Programmers can enter an exact rpm by specifying an *S-word*—in a program, the letter *S* followed by a number value denotes the exact desired rpm. Spindle speed is based on cutter material, cutter size, cutter type, and workpiece material.

Spindle Power

Spindle power relates to the available torque and horsepower requirements for various spindles. Spindle speeds operate within low and high spindle power ranges. Usually, these ranges exist on large machining centers. The low range permits heavy machining to occur without overloading the spindle. Machining centers switch to the required range automatically.

Feed Rate

The feed rate on a machining center is almost always expressed in feed per minute. The feed rate is set using an *F-command*. In the inch mode, feed rate is inches per minute; in the metric mode, it is millimeters per minute. Programming practices allow a decimal point to be included with the feed rate word.

Coolants

A *coolant* is the fluid used during machining to reduce heat caused by friction and to remove chips. Both flood and mist delivery systems are used on machining centers. Coolant nozzles are adjusted prior to machining to direct the coolant to the desired locations. Flood coolant is a continuous flow of a liquid, which usually is soluble oil. Mist coolant is an air-fluid mixture that is blown at the desired location. The M8 command turns on flood coolant while the command M7 turns on mist coolant. M9 cancels coolant flow.

> ### Note
>
> Command codes mentioned in this chapter will be discussed and explained further in later chapters of this textbook.

Auto Tool Changers

Virtually all machining centers permit tools to be loaded automatically from a magazine or turret. See **Figure 4-14**. Machining centers may have a magazine that can hold anywhere from 16 to 100 tools. The four most popular tool storage magazines systems are turret head, carousel, 180° rotation system, and pivot insertion system.

Turret

A turret is ordinarily found on older NC drill and tapping machines. A *turret*, or turret head, is a device made up of spindles where a number of tools can be mounted, **Figure 4-15**. A typical number of tools is six. Tools are loaded prior to program execution and the turret indexes to the various stations to select the desired tool. Tools are not removed during the tool change, which results in fast tool changes. The major disadvantage is the limited number of tools available for a program. Programs containing many tools could not be efficiently run using this tool change system.

Figure 4-14. A rotary tool magazine on a vertical machining center. This magazine holds 24 tools, but some models can hold as many as 100.

Carousel

Carousel storage systems are found on vertical machining centers. Tools are loaded in a coded drum called a *carousel*. The drum rotates with the desired tool to a tool change position. The vertical head with spindle moves down and picks up the tool. If a different tool is in the spindle, it will be put away on the drum prior to loading the desired tool.

180° Rotation System

In a 180° rotation system, a tool change command sends the spindle to a fixed tool change position, while at the same time the tool magazine is indexed to the proper position. See **Figure 4-16.** The tool changer rotates and grips the tool in the spindle and the tool in the magazine simultaneously. This system is used on either vertical or horizontal machines.

The *drawbar* (a device that grabs and pulls the tool holder retention knob) is removed from the tool holder in the spindle. The tool changer removes both tools before rotating 180° to replace the tool that was in the spindle with the new tool from the magazine. While the tool changer is rotating, the magazine repositions to accept the old tool. The tool changer then puts away the old tool and installs the new tool at the same time. The change is completed when the tool changer rotates back to its original position, ready for the next tool change command.

Figure 4-15. A tool turret used on a vertical machining center. The turret has capacity for six tools, but no tools are presently mounted on these spindles.

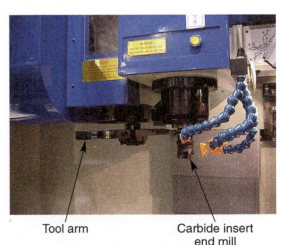

Figure 4-16. A 180° rotation system tool changer on a vertical machining center. Here, a tool is being selected to be put back into the magazine.

Pivot Insertion System

The pivot insertion system, **Figure 4-17**, has the same physical design as that of the 180° rotation system. Upon receiving a tool change command, the spindle moves to the tool change position. The tool magazine rotates the new tool to the proper position for the tool changer to remove it from its pocket. The tool changer removes the new tool from the tool magazine (located on the side of the machine), then pivots to the front of the machine where it grips and removes the toolholder from the spindle. The changer rotates 180° and places the new toolholder in the spindle. During this procedure, the tool magazine has indexed to the proper position to receive the old tool. The tool changer pivots back to the side of the machine and inserts the old toolholder in its pocket in the tool magazine. Finally, the changer returns to its home position and the CNC program continues. Although this system is slower than the 180° rotation system, its advantage is it protects the tools and system from possible chip damage.

Pallet Changers

Horizontal machining centers almost always use palletizing systems; such systems are now becoming more popular with vertical machining centers. See **Figure 4-18**. A system with two or more pallets allows machining of a workpiece mounted on one pallet to continue while a new workpiece is loaded on another pallet.

Part weight and shape determines the type of pallet that should be used. Large castings and weldments that require extensive setup time are candidates for use with pallet loading. Pallets are manufactured in a variety of sizes and shapes including round, square, and rectangular. Fixtures and vises are often mounted to pallets.

Tool magazine

Tool changer

Figure 4-17. This is a tool changer using the pivot insertion system on a horizontal machining center.

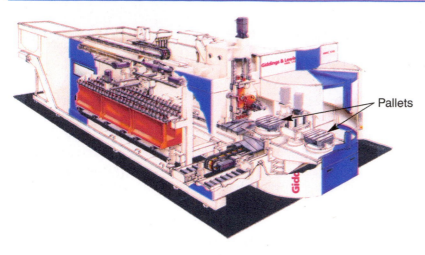

Figure 4-18. Horizontal machining center with a dual pallet system. Dual pallets allow the loading, unloading, or setting up of one pallet while the workpiece on the other pallet is being machined. (Giddings & Lewis)

Pallets

Rotating Devices

Rotating devices, such as rotary tables and indexers, are used on horizontal and vertical machining centers. Vertical machining centers usually have these devices mounted on their tables. Horizontal machining centers may have them built into the table itself. Rotating devices enable several sides of the workpiece to be machined during one setup. The advantages of this are that the number of setups can be reduced, there are fewer programs needed, parts can be produced faster, and it is easier to maintain accuracy between surfaces, since there is less handling of the workpiece.

Indexers are devices that allow the workpiece to be quickly rotated a specified number of angular degrees, **Figure 4-19**. Rotation speed is fast and cannot be controlled. Common devices include 90°, 45°, 5°, and 1° indexers.

Rotary tables are much more flexible than indexers. The rotation angle can be controlled more precisely, **Figure 4-20**. The table can rotate either clockwise or counterclockwise. The travel rate (rotation speed) can be controlled. The rotation angle can be set to within 0.001°, which results in 360,000 positions. The rotation of a rotary table is regarded as a machine axis. On horizontal machines, it is usually the B axis; on vertical machines, it is usually the A axis.

Dwell

Dwell is a command used by the machining center to pause tool movement at a particular time in a part program. It is usually used to relieve tool pressure caused by plunging, which is a characteristic of spotfacing, counterboring, or pocket machining. Pausing the tool briefly while spotfacing or counterboring results in a better finish. Dwell simply pauses axis motion; it does not affect other functions such as coolant flow or spindle rotation. Although the G04 code activates dwell, different controls

T-slots

Figure 4-19. This coupled indexer with indexing in 5° increments has T-slots to fasten workpieces or fixtures.

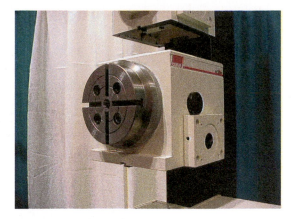

Figure 4-20. This rotary table can be rotated to within 0.001°, has a controllable rotation speed, and has 360,000 possible positions.

will use different addresses to specify time in seconds. The most popular address used is X. The actual command for dwell will be covered in the section dealing with G-codes.

Tool Length Offsets

Tool length offsets are values assigned to tools to allow for the differences in tool lengths. The object of using offsets is to treat the tools in a program as if they were the same length, when in reality they have various lengths.

On the machining center, tool length offsets are used for the Z axis. Offset values are entered by the operator by downloading them through a file during job setup or by using the keyboard to enter information into the controller (*manual data input* or MDI). The length of each tool is established by lowering a mounted tool or spindle face slowly until it lightly touches the top of a part or a setting block.

Tool length compensation is the term used when working with tool offsets. Tool length compensation permits the part programmer to ignore tool length as the program is written. Tool length is ordinarily measured from the tip of the tool to the nose or face of the spindle. However, depending on the toolholder style, this configuration may vary. With taper shank tooling there will always be a small gap between the nose of the spindle and the butt of the tool adapter flange. If tool lengths are measured away from the machine, this gap must be taken into consideration when determining tool length.

Some companies preset all cutting tools, resulting in consistency of length every time a particular tool is used. Some companies may use tool assembly drawings every time a particular tool is used. These drawings describe the cutting tool and give the setting length for each tool. With tool presetting, the operator must adjust the length of the tool to an exact value. Tool presetting is difficult and time consuming. It is usually used on horizontal machining centers.

Most CNC controls use three words related to tool length compensation. G43 is the G-code that calls for tool length compensation. There is one G43 code for each tool in the program. Included with the G43 code is an H-word that alerts the control which tool offset number is being used as the tool length. Good machine practice will require an H-word with the same value as the tool number to avoid mixups. For example, if tool one is being used, then H01 should be used as the offset number for that tool. Also included with the G43 command is a Z axis position departure. It is the location along the Z axis where the tip of the tool should stop prior to machining. In most applications it is set at .100″ above part zero. A typical program line may read:
N100G43H02Z.1

On other types of machining centers, such as the Mazak with a Mazatrol controller, the tool height offset is called up with the tool number. The control automatically accesses the tool file containing the offset value. Tool 11 offset is found in the tool file under the tool number 11.

Some controls may need a G-code to cancel tool length compensation. G49 is the code that cancels tool length compensation.

Setting Tool Length Offsets

There are three methods used to set tool length offsets. These methods are the plus direction trim, minus direction trim, and gage tool. The plus direction trim and minus direction trim tool length offset methods are covered in more detail in Chapter 9, when code G43 (tool length compensation) is covered.

Plus direction trim

This method involves touching the spindle face to part zero and setting Z fixture offset to this value. For example, if the axis readout shows a value of –12.050 in Z when the spindle face (spindle gage line) contacts the part surface (part zero), enter that value into the fixture offset Z file. The spindle is then fully retracted and a new tool installed. The tip of each tool used in the program is touched to the surface chosen as Z part zero and the axis readout for each tool is recorded in the offset files as a plus value.

Caution

Although the axis readout when touching the end of the tool is a minus value, the offset recorded is a plus value. This could prove to be a dangerous method to use if an individual failed to record offset as a positive (plus) value. This would result in a crash.

Minus direction trim

This method involves touching off the tip of each tool and entering, into the offset files, the Z minus value obtained from the axis readout. This means the Z value in the Z fixture offset file remains as a zero value. The Z offset value is the distance measured from the end of the tool with the spindle fully retracted to the Z axis home position (machine home). Z offset values will be negative values.

Gage tool

This method involves using the longest tool in your program set as the gage length tool. The tip of this tool is touched to the surface of the workpiece, which is called Z part zero. The Z value of this tool from machine home Z (full retract) to part zero is entered in the fixture offset file. The offset value for this tool will be zero. Each subsequent tool will be touched off to part zero and the tool length offset for each tool will be the difference between the length of that tool and the gage tool. The offset value of each tool other than the gage tool will be a negative value. Sometimes a setting block is used to set the tool lengths, rather than using the top finished surface of the workpiece. In this situation, the height of the setting block must enter into the calculation for each tool offset.

Note

Fixture offsets are distances in X, Y, and Z directions from home positions to the locations selected as part zero.

Tool Radius Offsets

Tool radius offsets are used to designate the size of the cutter being used in a program. A two digit D-word is used in a program to designate the tool offset. Offset values are manually entered in the MCU (machine control unit, or controller) to make adjustments to the cutting tool path. These offset values are based on the radius of the specific cutter being used. They are values stored in the controller in tool offset registers. For example, Fanuc machining centers generally can store 64 offsets. However, this capacity can be 199 offsets on some machines.

Usually, one offset is used per tool. In some applications, such as roughing and finishing, operations using end mills and shell end mills, two offsets may be used for the same tool. Offsets are used when cutting the periphery or contour of a workpiece. Using cutter offsets lets the programmer forget about the size of a cutter when determining cutter positioning. This will be explained in more detail when radius compensation G-codes (G40, G41 and G42) are discussed.

There are several reasons why offsets are used:

- **Various cutter sizes.** Offsets allow various cutter sizes to be used when machining. Resharpened cutters that are a different size than their original nominal size can be used in a program. If a cutter intended to be used is not available, a different size cutter can be used.

- **Workpiece sizing.** If the workpiece is cutting undersize or oversize in dimension because of a tooling problem, the workpiece can be machined to the correct print size by "lying" to the controller. Offsets can be set to smaller or larger values to affect workpiece sizes. For example, a dimension being machined that is a square island 2.000" × 2.000" is measuring at 2.002" × 2.002" because of cutter length deflection. Using the radius compensation feature of the machine, the offset value of the tool can be reduced 0.001" and the workpiece would be machined to the correct size.

- **Roughing and finishing cuts.** In a case where there is excess material, two or more offsets can be used for the same cutter. This allows the machine to take several passes along the workpiece. The first offset is a value larger than actual cutter size. This would cause the cutter to back away from the intended tool path, thereby removing a smaller amount of material. The tool path is then recut using a smaller offset value to remove additional material. This is repeated to remove excess stock until the final finish path is a radius offset value equal to the actual cutter size.

- **Tool wear adjustment.** As a tool dulls and cuts become smaller, the offset value may be reduced, causing the tool to cut to the correct size. However, dull tools can cause bad finishes and other machining problems. It is best to always use sharp cutters when machining.

- **Path calculations.** Not using cutter centerline programming allows the programmer to use radius compensation and actual print dimensions when computing the tool path. This makes tool path coordinates easier to calculate. An actual program line using offsets and radius compensation might appear as:
  N0128G42D13X5.5

Note

The program is telling the controller to adjust the tool path by moving the tool (location X 5.5") to the left of the tool path an amount found in offset D13. More detailed explanation will be given in later chapters.

Summary

Vertical machining centers are the most common type of milling machine used. Horizontal machining centers are very versatile and are becoming more popular. They have the capability of machining several sides of a workpiece.

Vises are the most common form of holding device. They are frequently grouped together for multiple part setups. Fixtures are specially designed-and-built holding and locating devices. They are used on both vertical and horizontal machining centers.

Spindle speeds and feed rates can be increased or decreased on machining centers by changing code values or using override switches. The carousel, 180° system, and pivot insertion system are the most widely used automatic tool changers with machining centers. The use of pallets on machining centers permits part setups to take place or job changes to occur while machining is being performed.

Rotary and indexing devices provide additional axis movement on a machining center. Different tool lengths and tool sizes are controlled on a machining center through the use of offset values loaded into computer files and accessed through program codes.

Chapter Review

Answer the following questions. Write your answers on a separate sheet of paper.

1. Name the three most common axis moves on a machining center.
2. Name the two types of horizontal machining centers.
3. An L-shaped holding device is called a(n) _____.
4. The holding system using specialty components mounted on a special plate is referred to as _____.
5. The S-word programs _____.
6. The major disadvantage of a turret system of tool changing is _____.
7. Why is dwell used on a machining center?
8. Explain the difference between a rotary table and an indexer.
9. A tool length value is accessed by using the letter _____ along with a number in a part program.
10. Touching the spindle face to part zero is part of the procedure named _____ when setting tool length offsets.
11. List three reasons for using tool radius offsets.

Activities

1. Attend a tool show and take photographs of various machining centers with different setups. Share this information with your classmates.
2. Use at least two methods of setting tool length offsets when performing setups on a machining center. Record all information and prepare a report on procedures.

Sufficient coolant is needed to prevent damage to the tool, reduce workpiece thermal deformation, improve surface finish, and flush away chips from the cutting area.

Chapter 5

Machining Center Tools, Inserts, Speeds, and Feeds

Objectives

Information in this chapter will enable you to:

- List various cutting tool materials.
- Classify the various grades of inserts according to their uses.
- Identify the types and applications of various drills used on CNC machining centers.
- Identify various taps and state their applications.
- List the purposes of using various cutting tool materials.
- Define hole operations such as drilling, tapping, reaming, countersinking, counterboring, and boring.
- Explain the uses of various types of milling cutters.
- State the difference between climb milling and conventional milling.
- Define the term *machinability*.
- Define the term *cutting speed*.
- List various factors affecting cutting speed.
- Define the term *feed*.
- Calculate rpm using a formula.
- List various factors that affect feed.
- Calculate feed rates using a formula.

Technical Terms

ball-nose end mills	face milling	root mean square
boring	face mills	roughing end mills
carbide	feed	shell end mills
center drill	fluteless taps	shell reamers
ceramic tools	high-helix drills	spade drills
cermets	high-speed steel (HSS)	spiral flute taps
chucking reamers	indexable insert drills	spiral pointed taps
climb milling	low-helix drills	spot drills
conventional milling	machinability	stellite
core drills	oil hole drills	step drills
corner radius end mills	plug taps	tapping
counterboring	profiling	toolholders
countersinking	reaming	twist drills
cutting speed		

Cutting Tool Materials

In machining, the intent is to remove material rapidly and economically, produce the best surface finish possible, and obtain the longest tool life. There are many cutting tool materials used with various types of coolants. Knowledge and selection of tool material is as important as the development of a CNC program. Cutting tools are manufactured from such materials as high-carbon steel, high-speed steel, cast alloys, carbides, ceramic tools, diamond tools and advanced cutting tool materials such as cermet, polycrystalline diamond (PCD), and polycrystalline cubic boron nitride (PCBN).

High-Carbon Steel

High-carbon steel is primarily used for machining wood or plastics and for low-speed operations, since it does not withstand temperatures above 350°F. High-carbon steels usually have a carbon content of 0.90%–1.5% carbon and various amounts of other elements.

High-Speed Steel

High-speed steel (HSS) is an alloy steel that has the ability to resist wear and to retain strength at high temperatures. It is used for nearly every type of rotating cutting tool, such as drills, reamers, taps, and end mills. High-speed steel tools are less expensive and tougher than carbide tools. They can be reused, because they can be resharpened easily and inexpensively. However, tools made from high-speed steel do not stand up to high temperatures and can dull easily.

HSS tools are used for machining both metallic and nonmetallic materials. Different types of high-speed steels include tungsten-based, molybdenum-based, and molybdenum-tungsten-based tool steels. Varying amounts of chromium, silicon, cobalt, and vanadium comprise the other ingredients found in these steels. HSS tools retain their hardness at much higher temperatures than high-carbon steel.

Cast Alloys (Stellite)

Stellite tools are cast alloys made up of approximately 50% cobalt, 30% chromium, 18% tungsten, and 2% carbon. These tools resist heat and abrasion. They remain tough at temperatures up to 1500°F.

Carbide

Carbide is a term commonly used to refer to cemented carbides. Carbide tools may be tungsten carbide, titanium carbide, or tantalum carbide and cobalt in different combinations. Carbide tools usually consist of indexable carbide inserts, although some tools still use brazed inserts. The advantages of carbide over HSS are that it has longer tool life under high temperature conditions and cuts hardened material better. Higher cutting speeds can be used with carbides. The material is brittle, however, and can fracture easily under interrupted cutting conditions.

There is a standardized system for identification of the various grades of carbide inserts. The system organizes insert grades according to their application or uses. The standard grades are as follows:

- **C1.** Roughing (cast iron and nonferrous materials)
- **C2.** General purpose (cast iron and nonferrous materials)
- **C3.** Finishing (cast iron and nonferrous materials)
- **C4.** Precision boring (cast iron and nonferrous materials)
- **C5.** Roughing (steel)
- **C6.** General purpose (steel)
- **C7.** Finishing (steel)
- **C8.** Precision boring (steel)

Uncoated carbides

Uncoated carbide cutting tools allow machining at speeds 3–4 times faster than high-speed steels. Tungsten is the major ingredient of this carbide, along with carbon. Powdered materials are heated, then pressed to form various shaped tool inserts. There are two primary categories of carbide inserts. One category consists of straight tungsten carbides, which are hard and possess very good wear resistance. They are used for machining cast irons and nonferrous materials. The other category includes a mixture of tungsten and tantalum or titanium carbide or a combination of all three. Its best use is for machining steel.

Coated carbides

Higher metal-removal rates can be accomplished through higher cutting speeds and higher feeds. Longer tool life is attained through using one of two methods used to coat carbide inserts as well as other cutting tool materials. These methods are chemical vapor deposition (CVD) and physical vapor deposition (PVD). CVD is a process that transforms gaseous molecules into a solid material in the form of thin film or powder on the surface of a substrate. PVD is a vacuum coating process that deposits substances on nonmetals to increase the life and productivity of cutting tools. Coatings used are titanium carbon (TiC), titanium carbo nitride (TiCN), titanium nitride (TIN), and titanium aluminium nitride (TiAlN).

Ceramic

Ceramic tools are made primarily from aluminum oxide, sometimes with titanium, magnesium, or chromium oxides added. Ceramics are quite brittle and require very rigid tooling and setups. Ceramic tooling should be used when carbide tooling shows wear.

Ceramic tools are shaped into triangular, square, or rectangular inserts. Ceramic tools are very hard and wear resistant and are used on rigid and powerful machine tools. They are used on hardened materials with speeds about twice that of carbide tooling. Two methods of coating inserts and other cutting tools are used. CVD and PVD are the methods used to provide cobalt enrichment to inserts.

Diamond Tools

Diamond-tipped cutting tools are used on nonferrous and nonmetallic materials requiring close tolerances and very high surface finishes. They are usually applied to finishing operations with cutting speeds up to 15 times that of other cutting tool materials.

Cermets

Cermets are ceramic/metal composites comprised mainly of titanium carbide (TiC) and titanium nitride (TiN). They possess moderate to high toughness with excellent edge wear resistance. Cermets are used for medium- to high-speed machining on carbon steels and alloys, stainless steels, and cast steels.

Polycrystalline Diamond (PCD)

Polycrystalline diamond (PCD) is a thin film diamond tip brazed onto a standard carbide insert. It is used for close-tolerance work requiring superior surface finish. PCD provides long tool life when milling highly abrasive nonmetals and nonferrous metals.

Polycrystalline Cubic Boron Nitride (PCBN)

Polycrystalline cubic boron nitride (PCBN) is used to process hardened ferrous metals used for high-speed machining. PCBN hardened tools possess high thermal and mechanical shock resistance, and have excellent wear characteristics.

Drilling

Drilling is the operation of producing a hole. Drills are available in different styles, diameters, and materials.

Twist Drills

Twist drills are the most common type of drill and are available with either straight or tapered shanks. They may be high-speed steel, carbide, carbide-tipped, or have carbide inserts. Different materials may be used for different types of drills, such as low-helix and high-helix drills. Drill point angles on HSS drills are different for various materials—the angles increase as the material gets harder. Mild steel is usually drilled with a 118° included drill point angle, like the one shown in **Figure 5-1**. Stainless steel is usually machined with a 135° drill point. Drills are manufactured in fractional, letter, number, and millimeter sizes.

Center Drills

Center drills, also known as *combination drill and countersink tools*, are used to produce accurate starter holes so that drills will begin in perfect alignment and location. See **Figure 5-2**. Hole location accuracy is greatly increased with the use of center drills, especially on rough surfaces. Center drills vary in diameter, with a No. 1 being the smallest size. The center drill should be machined to a depth where the conical portion of the center drill is slightly larger than the drill to be used to size the hole.

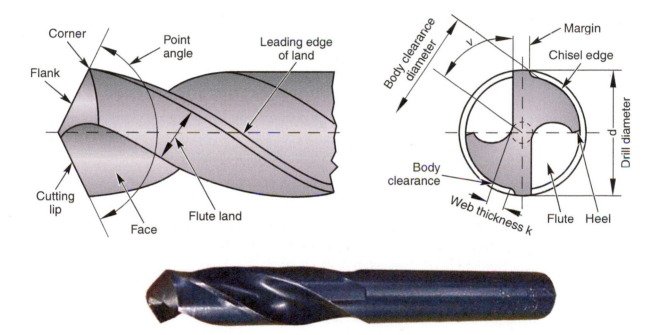

Figure 5-1. Twist drills normally have 118° included angles, usually are made from high-speed steel, and come in fractional, decimal, number, letter, and metric sizes.

Figure 5-2. This carbide center drill is a plain-type 60° included angle combination drill and countersink. It is used in cast iron, hardened tool steel, plastics, and steel with work-hardened surfaces. (Morse)

Spot Drills

Spot drills are short drills used to perform the same function as center drills. See **Figure 5-3**. They are used to accurately locate the position of a hole prior to drilling. The advantage over a center drill is that the spot drill is less likely to break.

Spade Drills

To produce a large hole, a twist drill needs to be quite large, and thus very expensive. *Spade drills*, **Figure 5-4**, are used to machine these larger holes without requiring a large diameter drill. With replaceable carbide inserts used on spade drills, only the insert (blade) is changed to drill a different size hole. The length of the tool does not change, which eliminates the need to reset tool length.

Core Drills

Core drills are three- or four-flute drills used to enlarge existing cored, drilled, or punched holes. See **Figure 5-5**. Drills with three or four flutes provide a better finish than two-flute drills.

Figure 5-3. The spot drill is used for a more accurate and faster spotting location for follow-up drilling. The design eliminates wandering of the drill. (Morse)

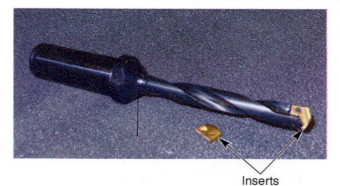

Inserts

Figure 5-4. This spade drill has replaceable carbide inserts and coolant holes. Changing inserts permits an economical way of drilling large holes.

Figure 5-5. A straight shank core drill is used to enlarge and straighten cored or drilled holes. (Morse)

Oil Hole Drills

Oil hole drills are twist drills with one or two oil holes running from the shank end of the drill to the cutting point. See **Figure 5-6**. Oil or cutting fluid is fed through these holes for chip flushing and drill point cooling. These drills are used for deep holes.

High-helix Drills

High-helix drills have a high angle (35°–40°) on the flutes, **Figure 5-7**. This angle assists in removing chips more rapidly when machining deep holes. These drills are used in aluminum, copper, die-cast, and other materials that typically cause chips to clog in the hole.

Low-helix Drills

Low-helix drills have a low angle on the flutes, **Figure 5-8**. The low angle prevents the drill from digging into soft material, such as brass, bronze, copper, and many plastics.

Step Drills

Step drills, also called *Subland drills*, provide multiple drilling operations in one tool. See **Figure 5-9**. These tools are specially designed and sharpened for specific hole configurations. They save in machining time and tool changes. However, several tools should be in stock in case of tool breakage.

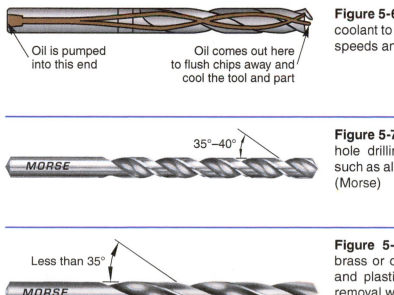

Oil is pumped into this end

Oil comes out here to flush chips away and cool the tool and part

Figure 5-6. Holes in the body of the drill deliver coolant to the point, permitting higher feeds and speeds and reducing temperatures.

35°–40°

MORSE

Figure 5-7. The high-helix drill is used for deep hole drilling in low-tensile-strength materials, such as aluminum, magnesium, and zinc alloys. (Morse)

Less than 35°

MORSE

Figure 5-8. Low-helix drills are used to drill brass or other readily machined copper alloys and plastics. These drills provide faster chip removal with less pull-in. (Morse)

Indexable Insert Drills

Indexable insert drills range in size from 5/8"–3". They are similar to spade drills, but use replaceable indexable carbide inserts. See **Figure 5-10**. Inserts used in these drills are manufactured in various shapes and sizes. These drills are used for drilling solid materials at rates 5 to 10 times that of twist drills. The disadvantages of these drills include higher horsepower needed, because of higher machining forces, and need for a high-pressure coolant system. Also, the carbide inserts are brittle and can break if subjected to shock. Improper use can cause chipping and cracking of inserts.

Tapping

Tapping is a machining process that produces threaded holes. It is a difficult operation because of the lubrication needed at the cutting edges and the required chip clearance. Difficulties also arise due to tough materials, dull taps, and deep threads. Lubrication is vital to thread quality and tap life.

Plug Taps

A *plug tap* is a type of hand tap that is frequently used in machine tapping on CNC machines. Taps are manufactured from materials such as carbon tool steel, high-speed steel, and carbide. Small taps (smaller than 1/4") are called machine screw taps; their major diameters are designated by numbers from 0 (0.060) to 12 (0.216). The two major categories of taps are NC (National Coarse) and NF (National Fine).

Figure 5-9. A step drill is frequently selected to minimize tool changes by using one tool to produce two or more hole diameters at the same location.

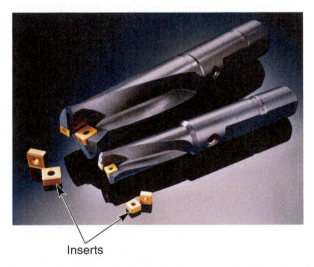

Inserts

Figure 5-10. Higher speeds and feed rates are possible with indexable insert drills. They eliminate the need for coolant, which poses a health hazard. (Kennametal)

Spiral Flute Tap

Spiral flute taps are used for tapping blind holes where the material is tough or stringy and chip removal is vital. See **Figure 5-11**. Helical flutes lift the chips out of the hole, preventing clogging, tap breakage, or thread damage.

Spiral Pointed Tap

Spiral pointed taps are used mostly for through holes, but can be used on blind holes with sufficient chip space at the bottom of the hole, **Figure 5-12**. Spiral pointed taps have several advantages:

*   Pushes chips forward, preventing the chips from packing in flutes or binding between the threads and work
*   Provides better flow of coolant or cutting fluid to cutting edges because flutes are larger
*   Requires less horsepower and torque, thereby reducing chances of tap breakage

Fluteless Taps

Fluteless taps are used to form a thread rather than cutting a thread, **Figure 5-13**. No chips are produced from this operation. They form threads in the hole with lobes on the outside edge of the tap. Tap drill sizes are larger when using this type of tap.

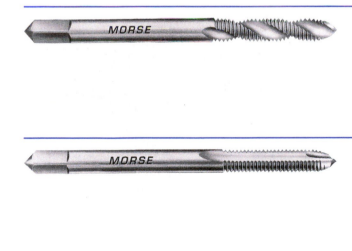

Figure 5-11. The spiral flute tap is a helical fluted tap used on ductile materials. Flutes cause shear-cutting action and permit passage of chips from the hole. (Morse)

Figure 5-12. Spiral point plug taps are designed for machine tapping in open or through holes. (Morse)

Figure 5-13. A fluteless tap is used to tap ductile metals where chip removal causes a problem. (Morse)

Reaming

Reaming is a finishing and sizing operation performed on a drilled or bored hole using a multiflute tool. Reaming will hold a tolerance of ±0.0002″ and produce a surface finish to 32 rms. The abbreviation rms stands for *root mean square*. It relates to surface roughness average and is used to describe surface finish produced by a metal cutting operation. For example, a fine torch cut will produce a surface finish of about 500 rms. Compare this to these common surfaces (the smaller the number, the smoother the finish):

Surface	Approximate rms
Magazine cover	16
Mirror/window glass	4
Cell phone surface	32

Reamers can have either straight or spiral flutes. Spiral-fluted reamers produce better finishes. Reaming does not correct hole straightness or location, but merely follows the previously machined hole.

Shell reamers are different size cartridges, usually 3/4″ or larger, that are mounted on either straight or taper-shank arbors, **Figure 5-14**. Various size heads will fit on the same size arbor. *Chucking reamers* are straight or taper-shank one-piece reamers with either straight or spiral flutes. See **Figure 5-15**.

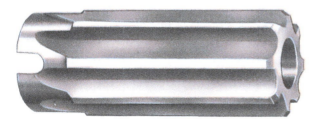

Figure 5-14. Shell reamers are interchangeable reamer heads used with separate arbors. (Morse)

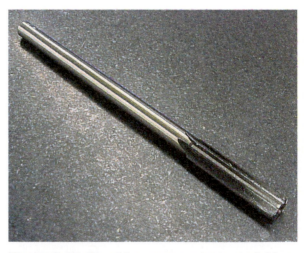

Figure 5-15. Chucking reamers have straight- or helical-flute shanks.

Countersinking

Countersinking is the operation of enlarging the upper end of a hole into a tapering, or conical, shape, **Figure 5-16**. The countersink angle is usually 82° or 90°, although some countersinks have a 45° or 60° included angle. The purposes for countersinking are to:

- Allow a flat-head screw to be flush or slightly beneath the surface when inserted into a hole.

- Deburr the upper end of a hole.

- Provide a bevel or chamfer for a threaded hole to protect the starting threads.

Counterboring

Counterboring is the operation of enlarging the upper end of a hole to produce a flat bottom portion so the head of a machine screw or bolt is below the surface. See **Figure 5-17**. Counterbores usually have two to eight

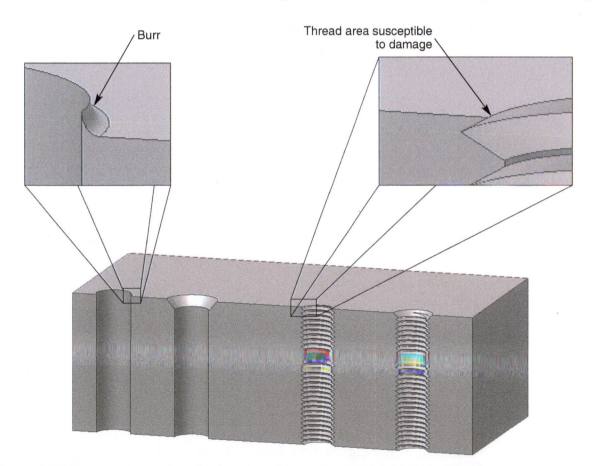

Figure 5-16. Countersinking tapers the upper portion of a drilled hole. Countersinks are used to deburr a hole and to protect starting threads.

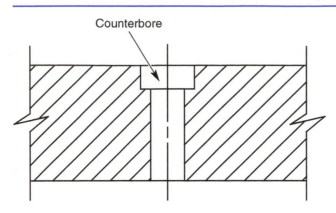

Counterbore

Figure 5-17. Counterboring is similar to countersinking, but produces a flat-bottomed enlargement of the upper end of a hole. This allows the head of a machine bolt or other fastener to be set below the surface.

flutes with either fixed or removable pilots. The pilot guides the counterbore into the hole, making sure it is concentric with the previous hole. It also aligns the tool 90° when counterboring on a rough or angular surface. Sometimes end mills are used to produce a counterbore. They are much cheaper then counterbore tools and can provide a better quality surface when used correctly. They also provide a greater capability to produce any size needed.

Boring

Boring is the operation of enlarging an existing hole with a single point cutting tool. It is used to straighten a hole and to make it concentric. It will also readjust the center location of a hole that is out of location. Boring is sometimes performed prior to reaming to make a straight hole. Boring will give a good finish and is sometimes done instead of reaming. There are many types of boring tools made from high-speed steel as well as carbide materials. The offset-type boring head is very popular with high-speed tooling, while the cartridge-type cutter using carbides is widely used in production machining and is more economical. The cartridge-type head has a single microadjustable cartridge for accurate size control.

Milling—Profiling

Profiling is the process of contouring a workpiece with a cutter that has teeth on its outside edge. *Face milling* is the process of creating a flat surface using an end mill or face milling cutter.

End Mills

End mills are used for profiling, slotting, facing, plunging, and cavity machining. See **Figure 5-18**. End mills are available in high-speed steel, solid carbide, insert carbide, and insert ceramic styles. End mills range in size from .032″–2.5″ diameters. They come with two, three, four, or more flutes. They also are available with indexable insert types that range in size from 0.5" diameter to 3.0" diameter. Larger sizes are referred to as face mills. End mills that produce male radii are known as *corner radius end mills*. End mills ground with a male radius on the end are known as *ball-nose end mills* and are used primarily for cavity work. End mills with grooves or scallops around the periphery are called *roughing end mills* and are used to remove excess material at a rate three times that of ordinary end mills.

Shell End Mills

Shell end mills are high-speed steel cutters mounted on special arbors, and range in size from 1.25"–6" in diameter. See **Figure 5-19**. They are used exclusively for face milling and have helical flutes.

Face Mills

Face mills are manufactured with high-speed steel, brazed carbide, and indexable carbide inserts. Face mills are available in sizes from 2" to more than 8" in diameter, **Figure 5-20**. Face mills with indexable carbides are the most popular for use in production situations.

Special Cutters

Special cutters are used for unique types of machining operations. These cutters include dovetail cutters, woodruff cutters, and T-slot cutters. See **Figure 5-21**.

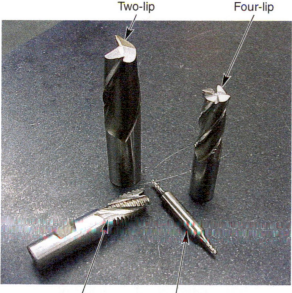

Two-lip   Four-lip

Roughing   Double-end
four-lip

Figure 5-18. Shown are several high-speed end mills.

Figure 5-19. A shell end mill is used with an adapter for face and slab milling in all types of materials. (Morse)

Figure 5-20. A face mill consists of a cutter body that is designed to hold multiple disposable carbide, steel, and ceramic inserts. (Kennametal)

Dovetail milling cutter   **Woodruff cutter with staggered teeth**   **T-slot cutter**

Figure 5-21. These various specialty cutters create unique shapes. (Morse)

Toolholders

Toolholders are used to secure cutting tools, such as drills, reamers, taps, end mills, face mills, and boring tools. Machining centers use a variety of styles and sizes of toolholders.

The most common taper used on larger machining centers is the No. 50 taper shank, which is a self-releasing taper. See **Figure 5-22.** The toolholder must have a flange or collar that the tool change arm can grip, **Figure 5-23.** It must also have a stud, tapped hole, or other device that holds the tool securely in the spindle. It is secured by means of a power drawbar or other holding mechanism. Common toolholders include the R8 collet, NMTB adapter, BT adapter, and V-flange adapter. See **Figure 5-24.**

Types of Feeds

The two types of feeds available when milling workpieces are conventional milling and climb milling. See **Figure 5-25.** The difference between conventional and climb milling is the direction in which the cutter is fed into the workpiece.

Taper Shank No.	Machine Type
#60	Very large machines
#50	Medium-size machines (20–50 horsepower)
#40	Small-size machines
#30	Very small machines

Figure 5-22. Taper shank numbers for use on various machine sizes.

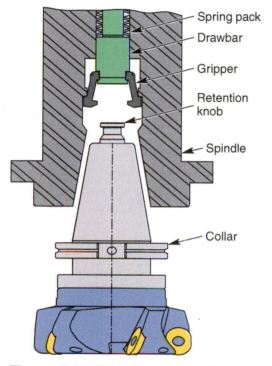

Figure 5-23. This illustration shows a drawbar with a gripper prepared to grasp the retention knob and pull the adapter into the spindle. (Kennametal)

R8 collect chuck, also called an M1TR taper

50 NMTB end mill adapter

50 BT dual flange type shank end mill adapter

50 taper dual flange type V-flange end mill adapter

Figure 5-24. These are the various toolholders available.

Conventional Milling

Conventional milling, or *up milling*, feeds the work against the rotation of the cutter. Cutting forces are directed upward. Conventional milling is most commonly used on manual or conventional milling machines. This method of cutting is used frequently on most ferrous materials, brass, and bronze. It is also used for roughing cuts on aluminum and its alloys.

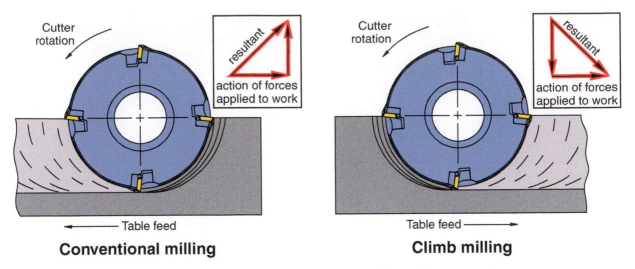

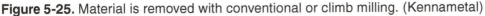

Figure 5-25. Material is removed with conventional or climb milling. (Kennametal)

Climb Milling

Climb milling, or *down milling*, feeds the work in the same direction as the rotation of the cutter. Climb milling is commonly used on CNC machining centers. Climb milling has cutter forces acting in a downward direction, which tends to pull the work into the cutter. Machines using this method of feeding must be equipped with a backlash eliminator to prevent this tendency to pull the work into the cutter. Climb milling works well for finishing aluminum and aluminum alloy products.

A major advantage of climb milling is that the work does not have to be clamped tightly because the cutter pushes the workpiece down, rather than lifting it as in conventional milling. This method also requires less power to machine. Finally, climb milling is good for milling deep, thin slots and thin workpieces.

Castings and hot rolled steel do not machine well with this method. The cutter teeth become damaged as they move in a downward direction through the outer scale on these materials.

Machinability

Machinability is the difficulty or ease with which a metal can be cut. Factors used in determining machinability include tool life, horsepower required, and resulting surface finish. Comparisons of various metals regarding machinability are based on AISI number B1112 steel having a rating of 100%. **Figure 5-26** gives the machinability ratings of various steels. Additional information regarding machinability can be found in *Machinery's Handbook*.

Machinability Ratings		
AISI number	Cutting speed (sfpm)	Machinability index (% Speed based on AISI B1112 as 100%)
B1112	165	100
C1120	135	81
C1140	120	72
C1008	110	66
C1020	120	72
C1030	115	70
C1040	105	64
C1060 (annealed)	85	51
C1090 (annealed)	70	42
3140 (annealed)	110	66
4140 (annealed)	110	66
5140 (annealed)	110	70
6120	95	57
8620	110	66
301 (stainless)	60	36
302 (stainless)	60	36
304 (stainless)	60	36
420 (stainless)	60	36

Figure 5-26. Cutting speed and machinability index for various steels.

Cutting Speed

Cutting speed for milling is determined by the distance that a point on the periphery, outer surface, of a cutter travels in one minute. The distance is measured in feet or meters. The cutter must be revolved a specific number of revolutions per minute. Cutting speed values are based on the hardness, machinability, and structure of the material being machined. Cutting speed is expressed as surface feet per minute (sfpm) or surface meters per minute (sm/m). Cutting speed values can be found in tooling manufacturers' catalogs, *Machinery's Handbook*, machine shop reference books, and speed/feed calculators.

Figure 5-27 shows the cutting speeds for carbide and high-speed steel cutters, based on various materials. The lower cutting speed is used for roughing and the higher cutting speed is used for finishing. To reduce the roughness of the surface finish, reduce the feed rate rather than increasing

Milling Cutting Speeds (sfpm)		
Material	**Carbide tool**	**High-speed steel tool (sfpm)**
Aluminum	1000–2000	500–1000
Brass	300–450	100–250
Bronze, soft	240–600	80–200
Bronze, hard	200–400	65–120
Cast iron, gray iron (soft)	125–200	60–80
Cast iron, ductile (hard)	150–250	50–70
Steel, free machining	250–350	125–175
Steel, mild (low carbon)	150–250	75–150
Steel, medium (0.3–0.6 carbon)	150–200	60–120
Steel, high carbon	100–175	50–90
Steel, 4140	100–150	50–70
Stainless steel	100–220	40–90
Magnesium	900–1600	400–1200
Titanium	125–175	25–70
Zinc die casting	1600–2000	800–1000

Figure 5-27. Milling cutting speeds for various types of metals.

the cutter speed. Using coolant will produce a better finish and lengthen tool life. Tooling manufacturers have very detailed cutting speed values for various materials, based on the grade of carbide used.

The proper combination of cutting speed with feed rate is based on several factors. These factors include the type of material to be machined, nature of heat treatment used, job setup rigidity, cutter setup rigidity, physical strength of the cutter, cutting tool material, spindle power, finish required, and cutting fluid used. Several of these factors affect primarily cutting speed, while some affect both cutting speed and feed rate. If the cutter is rotated too slowly, valuable machining time will be lost, while excessive speed can result in lost time due to the need to replace and regrind cutters.

Revolutions Per Minute (rpm)

The standard formula for calculating revolutions per minute is four times the cutting speed divided by the cutter diameter, or:

$$\text{rpm} = \frac{4 \times \text{CS}}{\text{D}}$$

where CS is cutting speed and D is cutter diameter.

For example, if a 4″ diameter high-speed steel cutter is to be used to machine aluminum at 1000 sfpm, the rpm is figured as follows:

$$rpm = \frac{4 \times CS}{D}$$

$$rpm = \frac{4 \times 1000}{4}$$

$$rpm = \frac{4000}{4}$$

$$rpm = 1000$$

If a 2.5″ carbide cutter is used to rough-machine mild steel at 150 sfpm, then the rpm is figured as follows:

$$rpm = \frac{4 \times CS}{D}$$

$$rpm = \frac{4 \times 150}{2.5}$$

$$rpm = \frac{600}{2.5}$$

$$rpm = 240$$

Feed

Feed is the distance that the work advances into the cutter. It is measured in inches per minute (ipm) or millimeters per minute (mpm). On most machines, feed is measured in inches per minute and is independent of the spindle speed. Feed rate can depend on factors such as depth of cut, width of cut, type of cutter, cutter material, cutter sharpness, workpiece material, strength and uniformity of the workpiece, required finish, required accuracy or tolerance, power and rigidity of the machine, and rigidity of the setup.

As the workpiece advances, each successive tooth of the cutter advances into it an equal amount, producing chips of equal thickness. Along with the number of teeth on the cutter and the spindle rpm, this chip thickness, or the feed per tooth (fpt), forms the basis for determining the feed rate. Chip thickness varies somewhat with the geometry of the cutter (positive rake, neutral rake, negative rake), but should be maintained within the range of 0.004″–0.008″. Chip thickness outside this range will result in either too little or too great a pressure on an insert for efficient machining. Too high a feed rate is indicated by excessive cutter wear. A cutter can also wear by taking too fine a feed. If the feed is too fine, a rubbing (instead of cutting) action may dull the cutting edge and generate excessive heat.

The data necessary to calculate feed rates is rpm, feed per tooth (see **Figure 5-28**), and the number of cutter teeth. The standard formula for feed rates is:

feed = N × fpt × rpm

where N is number of teeth, fpt is feed per tooth, and rpm is spindle speed in revolutions per minute.

Recommended Feed per Tooth for High-speed Steel Cutters (inches)

Material	Plain mills	Helical mills	Face mills	End mills	Form-relieved mills	Slitting saws
Low-carbon steel	0.005	0.008	0.010	0.005	0.003	0.003
Medium-carbon steel	0.005	0.008	0.009	0.004	0.003	0.002
High-carbon steel	0.003	0.005	0.006	0.002	0.002	0.002
Stainless steel	0.003	0.004	0.005	0.002	0.002	0.002
Soft cast iron	0.008	0.010	0.014	0.008	0.004	0.004
Medium cast iron	0.006	0.010	0.012	0.006	0.004	0.002
Malleable iron	0.006	0.010	0.012	0.006	0.004	0.002
Brass and bronze	0.008	0.010	0.013	0.006	0.004	0.003
Aluminum and alloys	0.010	0.012	0.020	0.010	0.007	0.004

Recommended Feed per Tooth for Carbide Insert Cutters (inches)

Material	Face mills	End mills
Low-carbon steel	0.013	0.007
Medium-carbon steel	0.012	0.006
High-carbon steel	0.012	0.006
Stainless steel	0.010	0.005
Soft cast iron	0.018	0.009
Medium cast iron	0.016	0.008
Malleable iron	0.014	0.007
Brass and bronze	0.012	0.006
Aluminum and alloys	0.020	0.010

Figure 5-28. These charts list recommended feed per tooth for high-speed steel cutters and carbide cutters.

The values in **Figure 5-28** are starting fpt values. Feed per tooth values for inserts are based on the different grades of inserts used and are expressed as a range of values. Most CNC machines have spindle speed and feed rate override controls. Using the spindle load meter, the operator can increase values to optimize machining. Increasing feed first, rather than speed, has the least effect on heat at the cutting zone and wear on the inserts. Adjustments to speed and feeds should be made in 10% increments while machining is taking place.

For example, to calculate the feed rate for finish machining brass with a four-flute HSS 3/4" end mill, first calculate the spindle rpm as follows:

$$rpm = \frac{4 \times CS}{D}$$

$$rpm = \frac{4 \times 250}{0.75}$$

$$rpm = \frac{1000}{0.75}$$

$$rpm = 1333 \ rpm$$

Next, calculate the feed rate:

$$feed = N \times fpt \times rpm$$

$$feed = 4 \times 0.006 \times 1333$$

$$feed = 31.99 \ ipm$$

To calculate the feed rate for rough machining soft cast iron with an 8-tooth 6" carbide face mill, first calculate the spindle rpm as follows:

$$rpm = \frac{4 \times CS}{D}$$

$$rpm = \frac{4 \times 125}{6}$$

$$rpm = \frac{500}{6}$$

$$rpm = 83 \ rpm$$

Next, calculate the feed rate:

$$feed = N \times fpt \times rpm$$

$$feed = 8 \times 0.018 \times 83$$

$$feed = 11.9 \ ipm$$

Depth of Cut

When a smooth finish is desired, both a rough cut and a finish cut should be used. A higher speed and a lighter feed are used for the finish cut. Roughing cuts should be deep, with a feed as heavy as the machine will safely permit. Heavy cuts should be taken with helical cutters that have as few teeth as possible. Cutters with only a few teeth are stronger than those with a larger number of teeth.

The first cut on a casting or forging should be made well below the surface skin to avoid dulling the cutter. Cuts less than 0.015″ in depth should be avoided. Light cuts and extremely fine feeds should be avoided. These cuts often cause the cutter to rub on the surface of the work, rather than bite into the material. This dulls the cutting edges. When performing end milling in low-carbon steel, the maximum recommended depth of cut is 1/3 to 1/2 of the cutter diameter. Approximately 0.020″ to 0.030″ should be left for the finishing cut.

Figure 5-29 shows feed and speed relationships when using cutter inserts. Many tooling manufacturers produce these charts.

Cutting Conditions Relating to Speed and Feed		
Condition	Speed	Feed
Rough cuts	⇓	⇑
Finishing cuts	⇑	⇓
End milling	⇑	⇓
Slotting	⇑	⇓
Hard material	⇓	⇒
Soft material	⇑	⇑
Scale	⇓	⇑
Increase tool life	⇓	⇑
Heavy depth of cut	⇓	⇓
Coolant	⇑	⇒

Increase ⇑
Decrease ⇓
Same ⇒

Figure 5-29. Speed and feed adjustments recommended for different conditions when using cutter inserts.

Summary

The majority of cutting tools are made from materials such as high-carbon steel, high-speed steel, Stellite, carbides, ceramics, diamonds, cermets, polycrystalline diamonds, and polycrystalline cubic boron nitrides. Insert grades are grouped into eight categories based on their application. Hole-producing operations include drilling, tapping, reaming, countersinking, counterboring, and boring. Profiling and face milling are the main operations performed when milling.

Machinability relates to the ease of machining a material. Cutting speed is the distance that a point on the edge of a cutter travels in one minute of time. Tables and charts exist that provide cutting speeds for various materials. Depth of cut varies according to whether rough or finishing cuts are being made.

Chapter Review

Answer the following questions. Write your answers on a separate sheet of paper.

1. What elements are found in Stellite?
2. What steel is used as the basic index for determining machinability?
3. Name the element that is the major ingredient of carbide.
4. What materials are included in the insert categories labeled C1–C4?
5. Identify the two methods used to coat carbides and other cutting tools.
6. What is the primary material used for ceramic tools?
7. List five types of drills available for producing holes.
8. Define the term *boring*.
9. What is the term for contouring a workpiece with an end mill?
10. List two advantages of climb milling.
11. Name three types of taps.
12. What function does a reamer perform?
13. List two kinds of carbides.
14. Why are high-helix drills used?
15. What operation enlarges the end of a hole to allow recessing the head of a bolt?
16. What is the included angle for a twist drill used on mild steel?
17. List five factors that affect feed rate.
18. What does the letter *N* pertain to in the formula used to calculate feed rate?
19. What causes cutter wear?
20. List three factors that affect the proper combination of cutting speed and feed.

Activities

1. Obtain feed and speed tables for several types of machining center tools.
2. Display all the types of tools used on machining centers.
3. Watch videos describing the various operations performed on machining centers.
4. Look at catalogs that describe machining center tooling in more detail.

5. Using the table, calculate the rpm for each of the machining situations listed in the table. Use the material in this chapter to obtain the information needed to calculate A through J.

Material machined	Milling operation	Tool material	Tool diameter	Speed (rpm)
Aluminum	Roughing	HSS	1/2"	A. _____
Magnesium	Finishing	HSS	1"	B. _____
Mild steel	Roughing	Carbide	2"	C. _____
Gray C.I.	Roughing	Carbide	4"	D. _____
Brass	Finishing	HHS	3/4"	E. _____
4140 steel	Roughing	Carbide	1 1/2"	F. _____
Stainless	Finishing	Carbide	1 1/2"	G. _____
Ductile C.I.	Finishing	Carbide	2"	H. _____
Aluminum	Finishing	Carbide	1 1/4"	I. _____
H.C. steel	Roughing	HSS	7/8"	J. _____

6. Using the table, calculate the rpm and the feed rate needed for the conditions provided. Use the material in this chapter to obtain the information needed to calculate A through J.

Material machining	Cutter used	Cutter material	No. of teeth	Feed per tooth	rpm	Feed rate
Low-carbon steel	6" helical mill	HSS	8	A. _____		
High-carbon steel	6" face mill	Carbide	8	B. _____		
Malleable iron	4" plain mill	HSS	8	C. _____		
Brass	4" slitting saw	HSS	24	D. _____		
Aluminum	8" face mill	HSS	12	E. _____		
Low-carbon steel	1" end mill	Carbide	4	F. _____		
High-carbon steel	¾" end mill	HSS	2	G. _____		
Malleable iron	6" helical mill	HSS	8	H. _____		
Brass	4" face mill	Carbide	10	I. _____		
Aluminum	4" face mill	Carbide	10	J. _____		

N10G20G99G40
N20G96S800M3
N30G50S4000
N40T0100M8
N50G00X3.35Z1.25T0101
N60G01X3.25F.002
N70G04X0.5
N80X3.35F.05
N90G00X5.0Z0T0101
01111
N10G20G99G40
N20G96S800M3
N30G50S4000
N40T0100M8
N50G00X3.35Z1.25T0101
N60G01X3.25F.002
N70G04X0.5
N80X3.35F.05

Chapter 6
Machining Center Carbide Insert Fundamentals

Objectives

Information in this chapter will enable you to:

- List several safety precautions to observe while machining with carbide inserts.
- List four milling cutter types used in machining.
- Identify different cutter mounting methods used for milling cutters.
- Differentiate between a negative and a positive insert entry angle.
- Differentiate between a 0° and a 45° lead angle.
- Recognize the different insert shapes.
- Explain the milling insert identification system.
- Select inserts for different machining situations.
- Explain the effect insert nose radii have on machining.
- Explain how the size of an insert is specified.

Technical Terms

adapter mounting	end mill adapter	lead angles
collet chuck adapter	mounting	milling
mounting	insert grades	nose radius
cutter pitch	insert size	Weldon shank

Milling

Milling is the process of removing material with a multitooth rotating cutter. Milling involves a nonrotating workpiece, interrupted tool engagement, varying chip thickness, and varying cutting forces. An operator should possess a working knowledge of cutter geometry, carbide grades, setup rigidity, and horsepower requirements. The four basic types of milling cutters are end mills, face mills, slotting cutters, and thread mills, **Figure 6-1.**

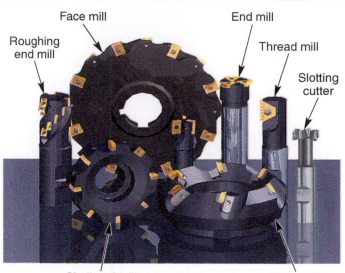

Roughing end mill

Face mill

End mill

Thread mill

Slotting cutter

Shell end mill

Face mill

Figure 6-1. Various holders using carbide inserts. (Kennametal)

Cutter Mounting

Milling cutters are mounted into the spindle of a CNC machine with arbors or adapters. Three commonly used methods of cutter mounting are adapter mounting, end mill adapter mounting, and collet chuck adapter mounting.

- **Adapter mounting.** With *adapter mounting*, the cutter is mounted on a piloted adapter that uses drive keys. The cutter is held in place by a lock screw or socket head cap screw. See **Figure 6-2**. This mounting is used for cutters six inches or less in diameter.

- **End mill adapter mounting.** In *end mill adapter mounting*, the cutter is held with a lock screw against a *Weldon shank*. This mounting is used for general-purpose work. See **Figure 6-3**.

- **Collet chuck adapter mounting.** With *collet chuck adapter mounting*, the collet reduces runout. This provides better part finishes and size control. See **Figure 6-4**.

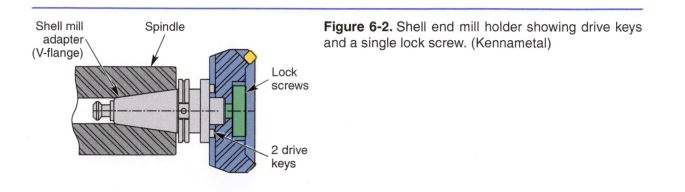

Shell mill adapter (V-flange)

Spindle

Lock screws

2 drive keys

Figure 6-2. Shell end mill holder showing drive keys and a single lock screw. (Kennametal)

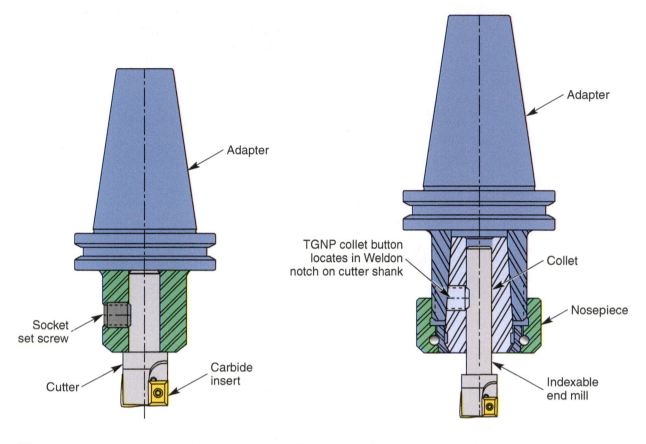

Figure 6-3. Various sizes of straight-shank end mills can be mounted in this style holder. (Kennametal)

Figure 6-4. Collet button provides positive drive for this collet chuck adapter, preventing slippage of the cutter shank. (Kennametal)

In addition to these three methods of holding cutters, there are many other types of holders for cutters. More information about these variations can be obtained from tool manufacturers' catalogs. If you are purchasing or using these types of tools, you should spend time familiarizing yourself with the many types of cutters and holders available.

Safety Precautions

Certain safety precautions should be taken when machining materials with carbide inserts. These precautions include:

- Protect against burns or physical injury that could result from high-temperature chips coming off the workpiece.
- Select an insert size and shape that is adequate for the job.
- Avoid continuous chips that could become entangled with the tool and work.

- Do not remove chips by hand because of their high temperature and sharp edges.

- Do not use an air hose to blow chips from the machine or your clothing.

- Keep tool overhang to a minimum to avoid tool chatter and possible insert breakage.

- Take extreme care to ensure both the tool and workpiece are tight and secure.

- Keep cutting fluids clean to avoid the possibility of foreign particles interfering with the workpiece finish.

- To avoid the possibility of igniting the coolant because of high temperatures generated during cutting, select the proper coolant and use sharp tooling.

- Do not overload the carbide inserts with excessive pressure; this could lead to insert breakage.

- Be careful when machining aluminum, magnesium, and titanium; they represent a potential fire hazard.

- Always use the largest size tooling available.

Cutter Selection

There are a number of factors to consider when selecting a cutter. These factors include cutter size, insert entry angle, milling cutter pitch, and lead angles.

Cutter Size

The size of the workpiece best determines the size of the cutter that should be used. A rule of thumb is that the cutter should be approximately one and one-half times the width of the workpiece. If the width is greater than any cutter size available, multiple passes should be made. With multiple passes, one-fourth the width of the cutter body should be outside the workpiece when making the first pass.

Insert Entry Angle

A negative entry angle is recommended to absorb the shock of the insert entering the cut. The strongest part of the insert is near the cutter body. A positive entry causes the tip of the insert to absorb the shock of entry. This could lead to the insert breaking.

Milling Cutter Pitch

Cutter pitch refers to the number of inserts found on a milling cutter. A coarse pitch is used for general-purpose milling where maximum depth is required and adequate horsepower is available. A medium pitch is used when it is necessary to have more than one insert in contact with the work. A fine pitch is used on workpieces that present an interrupted cut condition. Increasing the number of inserts on a cutter allows the feed rate (ipm) to be increased.

Lead Angles

Lead angles on a milling cutter have an effect on cutting force direction, chip thickness, and tool life. Lead angles can be 0°, 15°, 20°, and 45°. A 0° lead is used to produce a 90° shoulder. See **Figure 6-5**. For general milling applications where rigid conditions exist, 15° and 20° leads are used. A 45° lead angle allows higher feed rates to be used. As the lead angle increases from 0° to 45°, the amount of entry shock minimizes. The lead angle will affect chip thickness. **Figure 6-6** shows that as the lead angle changes, the chip thickness changes as well.

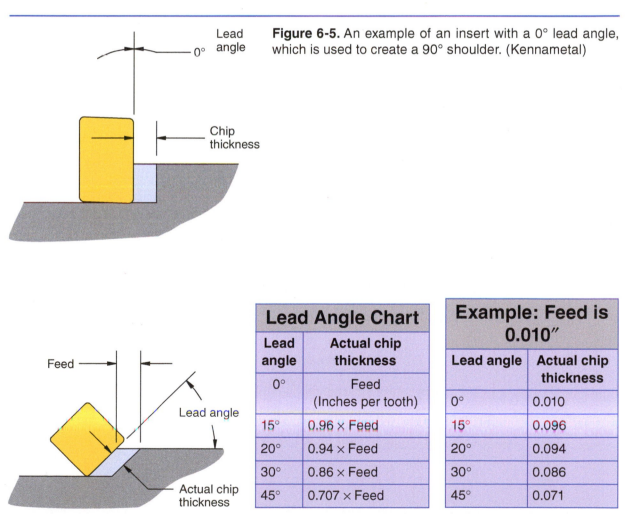

Figure 6-5. An example of an insert with a 0° lead angle, which is used to create a 90° shoulder. (Kennametal)

Lead Angle Chart

Lead angle	Actual chip thickness
0°	Feed (Inches per tooth)
15°	0.96 × Feed
20°	0.94 × Feed
30°	0.86 × Feed
45°	0.707 × Feed

Example: Feed is 0.010″

Lead angle	Actual chip thickness
0°	0.010
15°	0.096
20°	0.094
30°	0.086
45°	0.071

Figure 6-6. These charts show that a larger lead angle results in a thinner chip when the feed remains the same.

Insert Selection

There are a number of factors that help determine which insert to use. These factors include insert grade, insert shape, insert size, and nose radius.

Insert Grade

Carbide *insert grades* are based on their toughness and their ability to resist wear. A manufacturer's technical guide should be used to select the proper grade needed to machine different materials. Different grades are also based on the type of cuts being taken. Systems for grading carbide inserts based on the insert's application and its physical characteristics have been developed by the American National Standards Institute (ANSI) and the International Standards Organization (ISO).

Charts recommending the grades to be used for different machining conditions are available from manufacturers. There are also cross-reference charts available in catalogs to help make selections among different manufacturers' grades. In a case where a certain grade is not available from one manufacturer, these charts can be used to select a similar grade from another manufacturer.

Insert Shape

Different insert shapes accomplish specific types of machining operations on various workpiece shapes. See **Figure 6-7**. The strength of the insert also depends on the shape of the insert. For example, the round insert has the greatest strength and the most cutting edges, while the 35° diamond insert has the lowest strength. The triangular insert is very versatile for milling operations.

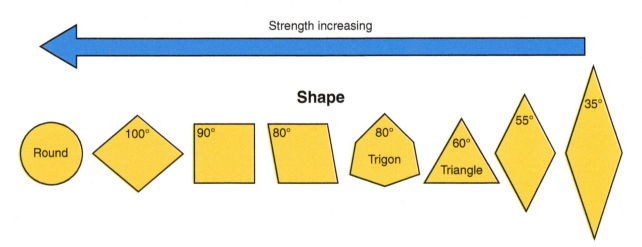

Figure 6-7. These are the basic carbide insert shapes. Strength of the insert increases as corners increase in angular value. The round insert is strongest. (Kennametal)

Insert Size

Insert size is determined by the largest inscribed circle (IC) that will fit inside the insert or touch all edges of the insert, **Figure 6-8**. The most common insert is a ¾″ square. As the size of the insert increases, the depth of cut can increase as well. A rule to follow is to set the depth of cut to no more than ⅔ the cutting edge length.

Nose Radius

The *nose radius*, **Figure 6-9**, will affect tool strength and surface finish. The larger the nose radius, the stronger the insert and the better the finish. However, if tooling and setup are not rigid, a large nose radius can cause chatter. The chart in **Figure 6-10** indicates the smoothness of surface finish that should result from various combinations of nose radii and feed rates.

Insert Identification System

A standardized letter and number system developed by ANSI, **Figure 6-11**, is used to identify and classify indexable carbide inserts. The identification reference is a combination of nine letters and numbers. The reference indicates the insert shape, clearance (relief angle), tolerance, insert type, size, thickness, point (nose) radius or chamfer, cutting edge condition, and (optional) manufacturer's identifier.

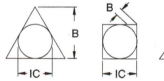

Figure 6-8. The diameter of an inscribed circle determines the size of the insert. In this installation, IC is the imaginary inscribed circle, B represents the distance from a tangent of the circle to the corner of the insert, and T is the thickness of the insert. (Kennametal)

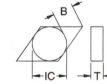
Nose radius

Figure 6-9. A large nose radius value results in a better finish, but can also cause chatter if tooling and setup are not rigid. (Kennametal)

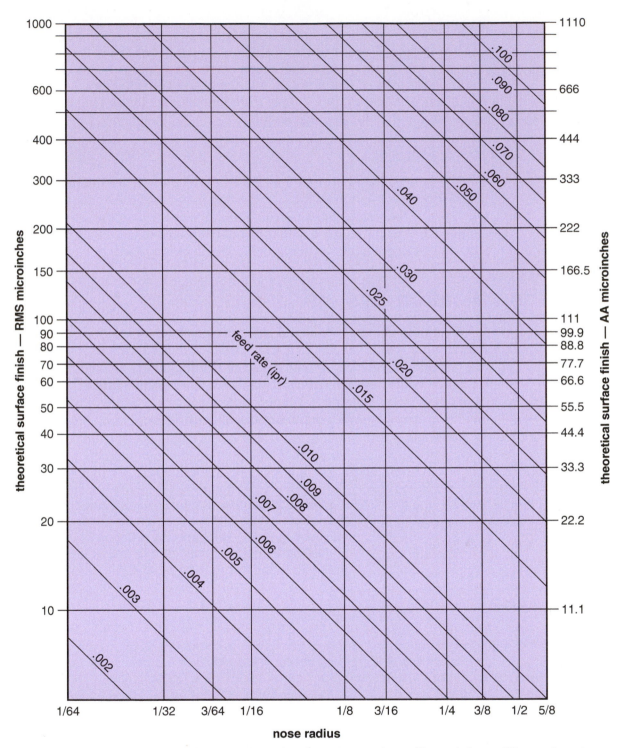

Figure 6-10. This metal finish chart predicts the smoothness that will result from different insert nose radius and feed rate combinations. For example, a 1/8″ radius and .010″ feed results in 25 rms finish, while a 1/32″ radius and .008″ feed results in 60 rms finish. (Kennametal)

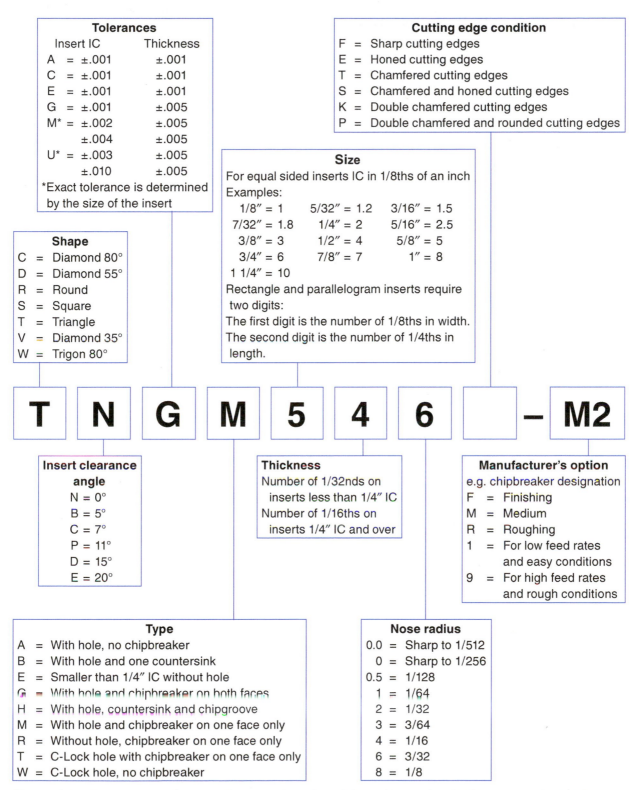

Figure 6-11. This graphic identifies the meaning of each letter or number in the standard identification system used to classify carbide inserts. The system was developed by the American National Standards Institute (ANSI).

Preparing for Machining

A machinist should be aware of the following items before starting a machining process.

- ☑ Check toolholders and replace any missing or damaged parts.
- ☑ Check the size of milling cutters, especially if reground.
- ☑ Check cutting edges of tooling for sharpness. Regrind if they are dull, nicked, or broken.
- ☑ Use the largest toolholder available for maximum rigidity.
- ☑ Use the shortest tool and toolholder for maximum rigidity.
- ☑ Use proper speeds and feeds for the type of tooling and the operation being performed.
- ☑ Use the correct cutting speed for the work material and tool material selected.
- ☑ Use a sufficient number of clamps and straps to fasten work to the table.
- ☑ Be sure the workpiece is sufficiently placed in a vise for maximum gripping power.
- ☑ Before making cuts, study the part drawing carefully to avoid mistakes.
- ☑ Set up clamps close to the workpiece and keep them parallel with the table.
- ☑ Always use parallels when holding the workpiece in a vise.
- ☑ Check for correct cutter rotation.
- ☑ Be sure all tools and the workpiece are clamped securely.
- ☑ Use the strongest inserts possible, depending on the type of cuts to be made.
- ☑ Use the maximum depth of cut and a coarse feed for roughing.
- ☑ Be sure the machine has adequate horsepower for roughing.
- ☑ Use negative rake inserts for greater strength.
- ☑ Use inserts with the largest nose radius for good finish and greater insert strength.
- ☑ Use climb milling for better finishes and longer tool life.
- ☑ Direct cutter force against the solid jaw of the vise.
- ☑ Check the cutter path to avoid striking clamps or fixtures.
- ☑ Remove all burrs and sharp edges from workpieces before fastening.
- ☑ Clean all tooling, fixturing, and workpieces before setting up.
- ☑ Use proper coolant to prolong tool life.

Summary

Milling is the process of removing material with a multitooth rotating cutter. There are four basic types of milling cutters. Cutters are mounted on adapters, end mill adapters, and collet chucks. Cutters are selected based on size, insert entry angle, insert pitch, and lead angles. Inserts are selected on the basis of grade, shape, size, and nose radius. A standard letter and number system is used to identify and classify cutting inserts.

Chapter Review

Answer the following questions. Write your answers on a separate sheet of paper.

1. How do you avoid the possibility of igniting coolant as a result of high temperatures generated from cutting?
2. List the four basic types of milling cutters.
3. What type of entry angle is recommended to absorb the shock of a carbide insert entering the machining cut?
4. What is *pitch*?
5. What milling cutter lead angle is best for allowing higher feed rates to be used?
6. List six shapes of inserts produced.
7. When dealing with inserts, what do the letters IC mean?
8. What do the letters CNMG mean when used to identify an insert?
9. What do the numbers 432 mean in the insert designation CNMG-432?
10. List five precautions that should be taken while preparing to machine a workpiece.

Activities

1. Obtain a sampling of the various types of inserts available based on shape and size, and arrange a display.
2. From a part drawing, select inserts best suited to manufacture the part.
3. Use an insert troubleshooting chart that shows causes and remedies for insert problems, and list ten problems that can arise from machining.
4. Obtain catalogs from manufacturers of cutting tool holders and inserts and develop a library of machining data.

In order to identify each of the many varieties of carbide inserts, part numbers are printed directly on the inserts. (Seco Tools AB)

```
N10G20G99G40
N20G96S800M3
N30G50S4000
N40T0100M8
N50G00X3.35Z1.25T0101
N60G01X3.25F.002
N70G04X0.5
N80X3.35F.05
N90G00X5.0Z0T0101
O1111
N10G20G99G40
N20G96S800M3
N30G50S4000
N40T0100M8
N50G00X3.35Z1.25T0101
N60G01X3.25F.002
N70G04X0.5
N80X3.35F.05
```

Chapter 7

Programming Process for Machining Centers

Objectives

Information in this chapter will enable you to:

- Apply the process for performing programming procedures.
- Determine the sequence of operations necessary to produce a workpiece.
- Determine a workholding method to use for machining a workpiece.
- Prepare a tool list for a programming job.
- Prepare a setup sheet for a programming job.
- Load a program into a machine using alternate methods.
- List the preferred procedure for performing program verification.

Technical Terms

direct numerical control (DNC)	manual data input (MDI)	setup sheet tool library
engineering drawing	program	tool list
flowchart	program	
inspection sheet	documentation	

CNC Documentation

Program documentation is the paperwork used for a CNC job. The amount of documentation developed or used by a company will vary. The documentation used by a programmer or machine operator may include the engineering drawing, setup sheet, tool list, tool library, program, and inspection sheet.

- The *engineering drawing* is a graphical description of the workpiece. It contains dimensions and tolerances. See **Figure 7-1**.

- A *setup sheet* shows how the part is held in a vise or fixture and where the holding device is located on the machine, **Figure 7-2**. It may also show where part zero is located, and may provide special instructions to the setup person and operator. Often, setup instructions are contained within the program itself.

133

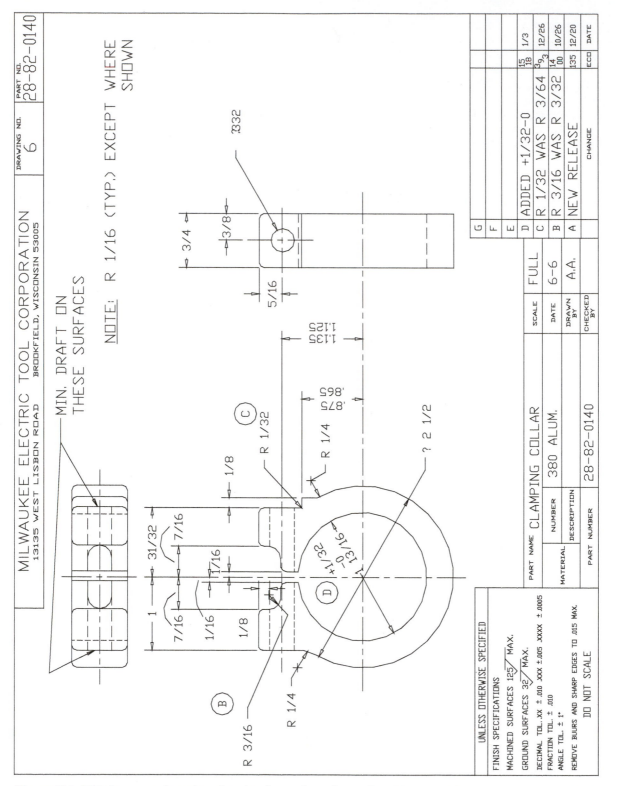

Figure 7-1. This is an engineering drawing for a clamping collar. Note that dimension changes from previous versions are listed.

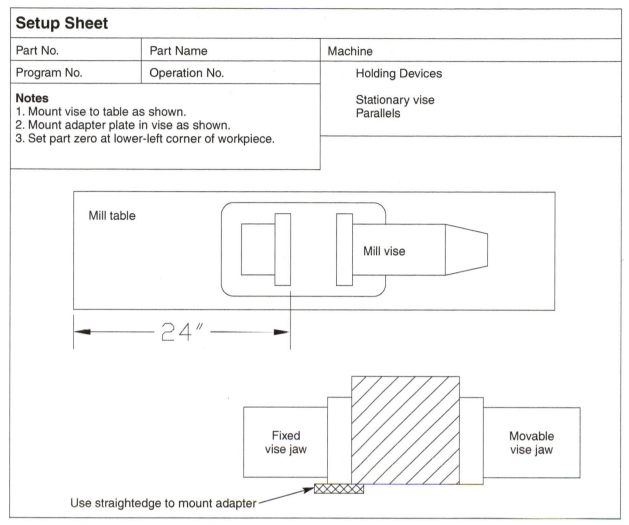

Figure 7-2. This setup sheet for an adapter plate describes activities that need to be performed to set up the job. These sheets often include sketches to aid in the setup of holding devices or fixtures. Setup sheets may also include information on where fixturing and tooling is located in the manufacturing facility.

- The *tool list* shows the tools used in the program and usually lists them in the sequence in which they are used. Tool lists may include tool holder identification numbers, tool lengths, insert identification, speeds, and feeds for each tool. Tool lists are usually designed by the company or programmer for internal use only. See **Figure 7-3**.

- The *tool library* is a catalog list of all tooling available for each machine in the shop, **Figure 7-4**.

- A *program* is a hard copy of the G-code program. See **Figure 7-5**.

GATEWAY CNC TOOL LIST AND OPERATION SHEET						
Part Name COVER	Part No. SME3		Course No. 444-381		Date 9-28-09	
Program No. 01333		Operation No. 10	Programmer RICH GIZELBACH			
Sequence No.	Tool No.	Pocket No.	Tool and Operation Description	Tool Dia.	Tool Offset	Tool Offset
1	T1	1	CENTERDRILL ALL HLS #3	#3 C'DRILL	H1	
2	T3	3	DRILL (6) 3/8 Ø HOLES	3/8	H3	
3	T4	4	C'BORE (6) 5/8 Ø HOLES	.625 EM	H4	
4	T8	2	DRILL (2) 13/64 HLS	13/64	H8	
5	T10	5	TAP (2) 1/4-20 HLS	1/4-20	H10	
6	T12	9	MILL (2) FLATS	1" EM	H12	
7	T6	7	MILL .375 SLOT	.375	H6	

Figure 7-3. A tool list form used to list the sequence of tool use, tool number in program, and location of the tool (pocket) in the machine. The type of tool and its description also are listed.

Pine Machine		CNC Vertical Mill Tool Library				
Tool No.	Tool Description	Tool Diameter	Gage Length	Comments		Insert No.
1	Spot drill	1/2	4.0			
2	Twist drill straight shank - HSS	3/16	6.3			
3	Spiral flute tap 1/4-28 UNF	1/4	5.2			
4	Twist drill straight shank - HSS	17/32	6.0			
5	Twist drill straight shank - HSS	21/64	6.7			
6	End mill 2-flute HSS	3/8	4.8	3/8" shank		
7	End mill 2-flute HSS	1/2	4.8	3/8" shank		
8	End mill 4-flute HSS	7/8	5.3	1/2" shank		
9	Twist drill straight shank - Carbide	1/4	5.2			
10	Twist drill straight shank - Carbide	29/64	7.2			
11	Spiral flute tap 1/2-20 UNF Class 2 H2	1/2	5.5			
12	Twist drill straight shank - HSS	19/32	6.1			
13	Twist drill straight shank - HSS	7/32	5.4			
14	Spiral flute tap 1/4-20 UNF Class 2 H2	1/4	4.9			
15	Spiral flute reamer straight shank - HSS	3/8	6.3			
16	Spiral flute reamer straight shank - HSS	5/8	5.9			
17	Kennametal 4" face mill	4	4.1	45° lead angle - 5 inserts		Sean-43

Figure 7-4. This is an example of a tool library. The tools that are available for use on a particular machine, in this case a CNC vertical milling machine, are listed here.

```
O0002
N5 G54 G90 S500 M03
N10 G00 X4 Y-2.875
N20 G43 H01 Z-.125
N30 G01 X-1 F3.0
N40 G91 G28 Z0
N50 G28 X0 Y0
N60 M30
```

This line tells the machine to move the tool rapidly in a straight line along two axes

Figure 7-5. An example of a portion of a G-code program.

- An *inspection sheet* provides the operator with information on the specific dimensions or features to check, tolerances on these dimensions or features, and the inspection devices or equipment that should be used. The form is designed with columns and rows to enter inspection results, **Figure 7-6.**

Production Part Approval Process

Customer/Vendor

Part Number

Page ___ of ___ Page ___

No.	B/P Dimension or Spec.	Inspection Method MMC	Inspection Results						Inspection Results						Acc	Rej
			1	2	3	4	5	6	1	2	3	4	5	6		

Job No.

Part Name

Comments

Comments

Signed Date

Title

Form #1210 Rev 6/7/2009

Figure 7-6. An inspection sheet lists the dimensions to be checked, the measuring tools needed to check these dimensions, and the results of the measurements taken.

Programming Sequence

Before writing a program, the programmer should understand the sequence of steps usually followed in the process of producing workpieces. This process will vary, depending on the size of the company. Small companies may have one or two people who handle the tasks in this process, while larger companies may distribute these steps among various departments.

Figure 7-7 is a *flowchart* showing the programming sequence used to manufacture a workpiece. The sequence shown can vary somewhat, since some tasks can be performed before others without affecting the final outcome.

Analyze the Print

Nearly all requirements necessary to produce the workpiece can be determined from analyzing the engineering drawing (usually referred to simply as the *print*). Items gathered from studying the print include holding devices needed, tooling, machine requirements, and part coordinate values. Prints that are not dimensioned properly often are marked by the programmer to provide Cartesian coordinate values. Tool positions are sometimes indicated on the print; coordinate tables are often used to define these tool positions.

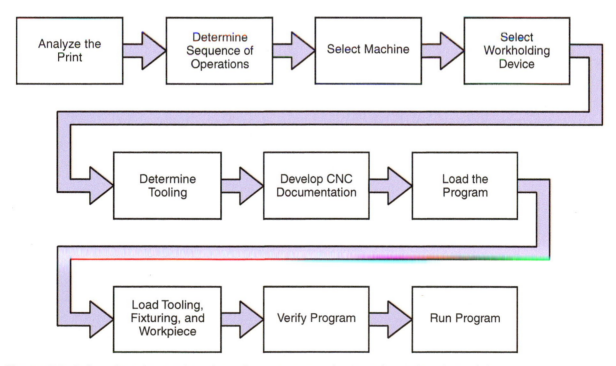

Figure 7-7. A flowchart is used to show the sequence of events in performing a job.

Determine Sequence of Operations

The types of machine operations necessary to produce the part must be determined, along with the sequencing of these operations. The shape of the workpiece may require several setups and several machines to complete the manufacture of the part. The order of operations—such as face milling, drilling, tapping, reaming, and end milling—and whether it is rough machining or finishing must be determined. This information provides the answers to the question "How do I manufacture this part?"

Select a Machine

Selection of equipment to machine the workpiece is based on many factors. These factors include the machine's availability, size of workpiece, travel limits, horsepower, operational cost, tool stations, number of workpieces needed, and the need for a horizontal or vertical spindle.

Choose a Workholding Method

The shape, size, and orientation of the workpiece determine the workholding method. The workholder could be simply a vise or multiple vises, or a complicated fixture requiring indexing or rotary movement. If a fixture is necessary, its design and manufacture are important additional steps in the programming process.

Identify Tooling Requirements

The necessary tools must be prepared and ordered for the job. In some cases, special tooling may be designed and manufactured to print specifications. Special tooling reduces the number of tool changes in a program because the tool performs multiple operations.

Load the Program

After the program is developed, it must be loaded into the memory of the CNC machine. Methods of loading the data can be grouped into the two categories: direct and indirect.

The *direct* method is called **manual data input (MDI)** and is performed at the machine, using the controller keyboard. This method is usually used for short programs. Loading long programs this way is inefficient and could tie up a machine for a long period. However, some machines permit entering a new program while they are executing or running another program.

The *indirect* method uses floppy disks, hard disks, punched tape, magnetic tape, or a hard-wire hookup. Programs are written in ASCII (American Standard Code for Information Interchange) format using a word processing software program and stored on a disk or other medium. Communication software then allows data to be transferred from the storage medium to the machine memory.

Personal computers can be connected to the machine to transfer a data (program) file. ***Direct numerical control (DNC)*** is a system using a computer that is hard-wired to several machines. The computer controls the operation of all machines in its network. Programs can be loaded from a shop office directly to a machine on the shop floor. Wireless communication between the computer and the CNC machine is the latest method used to load programs.

Load Tooling, Fixturing, and Workpiece

Machine setup includes the loading of all tools that are needed for the job (including the tool length measurements), the loading and alignment of any workholding devices (vises or fixtures). The location and setting of part zero is also part of machine setup. Adjusting of coolant nozzle locations may be part of machine setup, as well.

Program Verification

Before the program can be run, a number of verification steps should be completed. These steps identify mistakes in the program. The program verification process starts with a no-risk machine lock dry run to prevent any damage to equipment and injury to workers. The process continues until the program is run at full speed and a workpiece is created.

Machine lock dry run

This verification checks a program for syntax errors. After all setups are complete, the operator turns on the machine lock and dry run switches and buttons, increases feed rates, and watches the program run.

With machine lock activated, all commands in the program will be performed, except for actual axis movement. This procedure finds mistakes in the programs with the exception of correct tool paths. When no alarms occur, the program is ready to check for motion errors in the tool path.

Dry run

This verification checks for motion problems and is performed without loading a workpiece. The feed rate and rapid override switches are set to their lowest settings and the machine cycle is activated. The operator keeps a finger on the feed hold button to stop the machine if a motion problem becomes evident, such as the wrong direction or a possible crash with the holding device. The dry run may be attempted several times until all incorrect or inefficient movements have been corrected. Feed rate can be increased during the run with the override switches. Axis overtravels, offset, and compensation problems can be discovered during this verification. Rapid motions and feed motions are at the same rate.

Single block run without workpiece

This verification without a workpiece is performed to distinguish between rapid and feed moves. It is performed to prevent rapid moves into the part due to errors in G00 and G01 moves. Switches are still controlled by the operator, as in the dry run.

Single block run with workpiece

This is the last verification step and it is still performed using the override switches to control the speed of the program. Care should be taken when approaching the workpiece, so that no crashes occur. Castings can sometimes be a problem because of varying stock material amounts. Offsets can be adjusted to control size dimensions.

Automatic run

The program is run at full speed and the workpiece is checked for sizes and finishes. Adjustments can be made to conform to print requirements.

Note

Cad/Cam programs and modern CNC controllers now have the capability of verifying the tool path graphically.

Summary

There are several steps in processing a program. These steps include preparing CNC documentation; analyzing a print; determining the sequence of operations; selecting a machine to produce the workpiece; selecting a workholding method; determining tooling requirements; loading a program; and setting up the machine, tooling, and workpiece. The final step is to perform a program verification.

Chapter Review

Answer the following questions. Write your answers on a separate sheet of paper.

1. List three items that are included in program documentation.
2. What type of information is given on a setup sheet?
3. What information is found on a typical tool list?
4. What is a *tool library*?
5. Describe the difference between direct and indirect program loading.
6. List three methods of indirect program loading on a machining center.
7. How is the feed rate set in a dry run?
8. What is the purpose of a machine lock during a dry run?

Activities

1. Perform MDI loading of a program. Have your instructor supply a program to load.
2. Load a program into a machine using the proper media. Obtain a procedure sheet from your instructor.
3. Develop a setup sheet for the workpiece shown below. Obtain or devise a setup sheet for a machine in the shop.

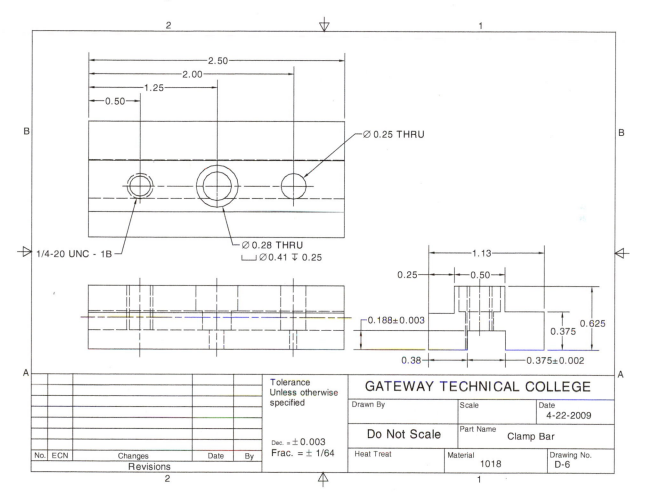

4. Develop a tool list for the workpiece in Activity 3.
5. Develop a sequence of operations for the workpiece in Activity 3. The workpiece was previously machined to 2.50 length, 1.13 width, and 0.625 height.

Tool carts ease the process of bringing multiple tools to the machining center. (Tibor Machine Products)

N10G20G99G40
N20G96S800M3
N30G50S4000
N40T0100M8
N50G00X3.35Z1.25T0101
N60G01X3.25F.002
N70G04X0.5
N80X3.35F.05
N90G00X5.0Z0T0101
01111
N10G20G99G40
N20G96S800M3
N30G50S4000
N40T0100M8
N50G00X3.35Z1.25T0101
N60G01X3.25F.002
N70G04X0.5
N80X3.35F.05

Chapter 8
Programming Codes for Machining Centers

Objectives

Information in this chapter will enable you to:

- Define the various alpha characters used to form words in a program.
- Describe preparatory codes (G-codes) used in a program.
- Define I- and J-words.
- Match various G-codes with their meaning.
- Explain the two methods of setting tool length compensation.
- Explain how fixture offset codes are determined.
- Define various miscellaneous codes (M-codes).

Technical Terms

block	IJK method	scaling
dwell command	initial plane	thread milling
fixture offsets	M-codes	variable block
G-codes	R-plane	word address format
home	radius method	words

Word Address Format

A *word address format* consists of *words*, which are a combination of alpha characters (letters) and numerical data. Alphabetic characters that are commonly used on machining centers include all the letters of the alphabet except *E* and *V*. Each of these 24 letters precede numerical data and define (identify) the purpose of the numerical data. Each word gives the CNC machine a single instruction or command. For example, both X5.000 and M05 are words representing single instructions.

A *block* is a group of words that tells the computer to act upon a complete statement of instructions. For example, N100 G00 X10.00 Y5.00 means this is line 100 (N100) of the program, and that the tool should rapid traverse (G00) to a position of 10″ in the X axis (X10.00) and 5″ in the Y axis (Y5.00). The *variable block* word address format allows a block of information

145

(instructions) to be any length. Words do not have to be in any particular order. However, it is good programming practice to sequence words in a definite order to allow a programmer to easily locate and read data in a program. The recommended letter sequence is the order presented in the following section.

Address Definitions

As previously mentioned, word addresses consist of letters followed by numerical data. The letters define the purpose of the numerical data. The following paragraphs describe the purpose of each letter:

- **O.** Identifies a program. It is the very first word in a program and is used in conjunction with a four-digit number from 0001 through 9999. For example, Program 12 is identified as O0012.

- **N.** Designates a sequence number, line number, or block number in a program. The letter *N* is used to identify a line in a program for organizational purposes and ease in editing. For example, Line 550 is designated as N550. Lines do not have to be in precise order, can be repeated, and may be omitted (to save memory).

- **G.** Identifies a preparatory function. Preparatory codes set various modes in a program. Although many G-codes exist, only a few are used in most programs. G-codes will be described in greater detail later in this chapter.

- **X.** Identifies a coordinate position along the X axis. Newer CNC machine controllers use this word with a decimal format, while older controllers rely on a specific number of digits. This will be explained in detail later in the chapter.

- **U.** Replaces the letter *X* when doing incremental moves.

- **Y.** Identifies a coordinate position along the Y axis.

- **W.** Replaces the letter *Y* when doing incremental moves.

- **Z.** Identifies a coordinate position along the Z axis.

- **A.** Designates rotation about the X axis. The rotation is most likely to be handled with a rotary device mounted to the CNC machining center.

- **B.** Designates rotation about the Y axis. This operation is usually used on horizontal machining centers. For example, the line N060 B30 tells the rotary table or device to rotate to a 30° position. On a horizontal machining center, the rotary device is mounted inside the machine as part of the table assembly. Some manufacturers may use the B-code for a rotary device on the vertical machining center.

- **C.** Designates rotation about the Z axis.

- **R.** This letter has two purposes: First, it designates the radius value in a circular movement. Second, it specifies the rapid plane height for a canned cycle command. This letter will be discussed in detail later in this chapter.

- **I.** This letter also has two purposes: First, it is used to specify the location of the arc center when making a circular move. The *I* specifies the distance and direction in the X axis. Examples will be given when covering G2 and G3 circular moves. Second, it specifies the amount of tool "back away" at the bottom of a fine boring cycle.

- **J.** Specifies the location of the arc center when making a circular move. The *J* refers to the distance and direction in the Y axis.

- **K.** Specifies the location of the arc center when making a circular move. The *K* refers to the distance and direction in the Z axis.

- **Q.** Specifies the peck depth for each pass in a peck drilling canned cycle.

- **P.** Specifies the length of time for dwell commands, which are used to pause axis motion for a specified period. It is used with a G4 code and is expressed in seconds or fractions of a second. No decimal points are allowed with the P-word. Thus, 1.5 seconds would be expressed as P1500 and 0.5 seconds as P500.

- **L.** Used with subroutines to specify the number of times that subroutine should be repeated. It is also used with canned cycles to designate the number of holes that should be machined.

- **F.** Specifies feed rate in inches per minute (ipm) using the decimal format. For example, F12 indicates 12 ipm and F7.5 indicates 7.5 ipm.

- **S.** Used to program a spindle speed. The format is S*xxxx*. For example, S1000 indicates a 1000 rpm spindle speed and S650 indicates a 650 rpm spindle speed.

- **T.** Used to select a tool station. The format is T*xx*. For example, T05 indicates Tool 5 and T1 indicates Tool 1. Note: In programming, leading zero suppression is used. This allows the zero to be dropped, so Tool 5 can be identified as either T05 or T5.

- **M.** Calls for a variety of miscellaneous functions. M-codes act like on/off switches that control systems such as spindle rotation and coolant flow. M-codes are covered later in this chapter.

- **D.** Specifies cutter radius compensation offset. The format is D*xx*. For example, D6 indicates a radius compensation offset of 6 and D07 indicates a radius compensation offset of 7.

- **H.** Specifies tool length compensation offset. The format is H*xx*. For example, H1 indicates a tool length offset of 1 and H08 indicates a tool length offset of 8.

G-Codes

G-codes are preparatory codes because they place the machine in a particular operating mode. The G-codes discussed in this textbook are standard codes recognized by EIA/ASCII standards. The G-code references will relate to Fanuc designations in regard to standard or optional features. The following terms are important to know when discussing G-codes.

- **Status.** Codes will be considered standard or optional with regard to Fanuc.
- **Initialized.** A code is active when the machine power is turned on.
- **Modal.** A code stays in effect until it is changed or cancelled.
- **Nonmodal.** Event occurs one time.

G-codes are also arranged by groups, which allow only one G-code within that group to be in effect. **Figure 8-1** illustrates the various G-codes according to group, function, status, and initialization.

G00—Rapid Traverse

The G00 code moves the table at maximum speed available for the machine. Common rates range from 400 ipm to over 1000 ipm. When used for positioning, movement along the axes is at same rate. Therefore, movement may not be in a straight line when using more than one axis. Sometimes separate moves are made to prevent collision with fixturing. The following program example is for the two-axis move illustrated in **Figure 8-2**.

```
O0001
N5G54G90S600M03
N10G00X2Y-2          Rapid to position
N15G43H01Z.1         Drill down
N20
⇓ (Program continues)
N25
```

G01—Straight Motion

The G01 code moves the tool in a straight line at a programmed feed rate (ipm) and motion. Both rate and motion remain at a modal condition. Operations include hole making, step milling, face milling, and angular milling. The following program example is for the two-axis move illustrated in **Figure 8-3**.

```
O0002
N5G54G90S500M03
N10G00X4Y-2.875      Rapid to Position 1
N20G43H01Z-.13       Rapid to depth
N30G01X-1F3.0        Feed to Position 2
N40G91G28Z0
N50G28X0Y0
N60M30
```

G-Code List – Fanuc

Code	Description	Group	Initialized	Modal
G00	Rapid positioning	A	Yes	Yes
G01	Linear cutting motion	A	No	Yes
G02	CW cutting motion	A	No	Yes
G03	CCW cutting motion	A	No	Yes
G04	Dwell	0	No	No
G09	Exact stop check	0	No	No
G10	Tool Offset Input by Program Command	0	No	No
G17	X-Y plane selection	B	No	No
G18	X-Z plane selection	B	No	No
G19	Y-Z plane selection	B	No	No
G20	Programming in inches	C	Yes	Yes
G21	Programming in mm	C	No	Yes
G22	Stored stroke check function on	0	No	Yes
G23	Stored stroke check function off	0	Yes	Yes
G27	Zero check return	0	No	No
G28	Return to home position	0	No	No
G29	Return from zero return	0	No	No
G30	2nd reference point return	0	No	No
G31	Skip function (used for probes and tool length measurement systems)	0	No	No
G33	Constant pitch threading	0	No	No
G34	Variable pitch threading	0	No	No
G40	Cutter radius compensation off	D	Yes	Yes
G41	Cutter radius compensation left	D	No	Yes
G42	Cutter radius compensation right	D	No	Yes
G43	Tool length offset positive	E	No	Yes
G44	Tool length offset negative	E	No	Yes
G45	Tool offset single increase	0	Yes	Yes
G46	Tool offset single decrease	0	Yes	Yes
G47	Tool offset double increase	0	No	Yes
G48	Tool offset double decrease	0	No	Yes
G49	Tool offset compensation cancel	E	Yes	Yes
G50	Scaling cancel	F	Yes	Yes
G51	Scaling on	F	No	Yes

(Continued)

Figure 8-1. This is the G-code list for Fanuc controllers. Parameters can be set to determine which codes are initialized upon startup.

Code	Description	Group	Initialized	Modal
G52	Local coordinate system selection	0	Yes	Yes
G53	Machine coordinate system	0	No	No
G54 to G59	Work coordinate systems	G	No	Yes
G59.1	Use preset work coordinate system 7	G	No	Yes
G59.2	Use preset work coordinate system 8	G	No	Yes
G59.3	Use preset work coordinate system 9	G	No	Yes
G60	Extended work coordinate systems	0	No	Yes
G61	Set path control mode: exact path	H	No	Yes
G61.1	Set path control mode: exact stop	H	No	Yes
G64	Set path control mode: continuous	H	No	Yes
G65	Custom macro call	0	No	No
G66	Custom macro modal call	I	No	Yes
G67	Cancel custom macro call	I	Yes	No
G68	Coordinate system rotation	0	No	Yes
G69	Cancel rotation	0	Yes	Yes
G73	High-speed drilling canned cycle	J	No	Yes
G74	Left-hand tapping canned cycle	J	No	Yes
G76	Fine boring canned cycle	J	No	Yes
G80	Cancel canned cycle	J	Yes	Yes
G81	Simple drilling cycle	J	No	Yes
G82	Drilling cycle with dwell	J	No	Yes
G83	Peck drilling cycle	J	No	Yes
G84	Canned cycle: right-hand tapping	J	No	Yes
G84.2	Direct right hand tapping canned cycle	J	No	Yes
G85	Canned cycle: boring, no dwell, feed out	J	No	Yes
G86	Canned cycle: boring, spindle stop, rapid out	J	No	Yes
G87	Canned cycle: back boring	J	No	Yes
G88	Canned cycle: boring, spindle stop, manual out	J	No	Yes
G89	Canned cycle: boring, dwell, feed out	J	No	Yes
G90	Absolute programming (type B and C systems)	K	No	Yes
G91	Incremental programming (type B and C systems)	K	No	Yes
G92	Programming of absolute zero point	0	No	Yes
G94	Units per minute feed rate mode	0	No	Yes
G95	Inch per minute/Inch per revolution feed (type A system)	0	No	Yes
G96	Constant cutting speed (constant surface speed)	0	No	Yes
G97	Constant rotation speed (constant RPM)	0	No	Yes
G98	Return to initial plane	L	Yes	Yes
G99	Return to rapid plane	L	No	Yes

Figure 8-1. *Continued.*

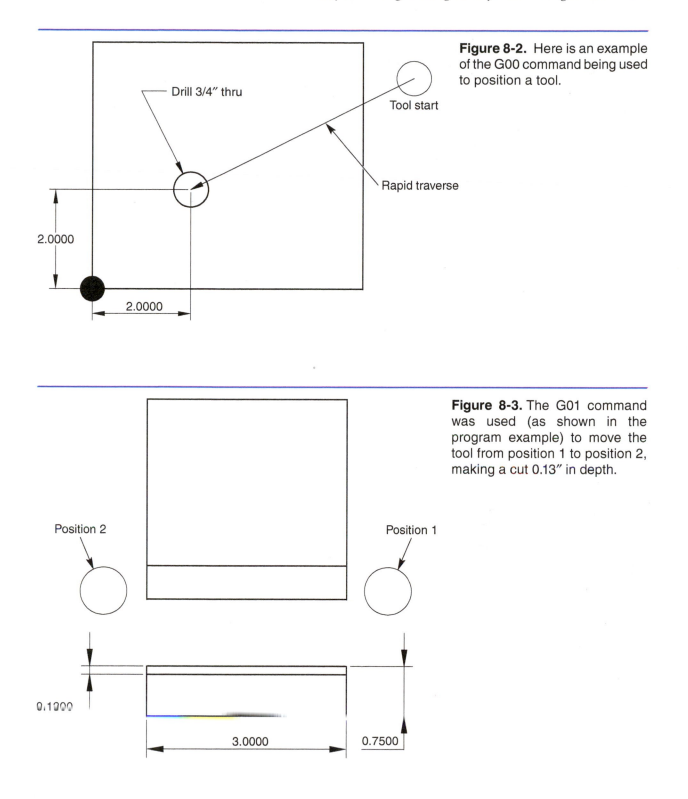

Figure 8-2. Here is an example of the G00 command being used to position a tool.

Drill 3/4″ thru

Tool start

Rapid traverse

2.0000

2.0000

Figure 8-3. The G01 command was used (as shown in the program example) to move the tool from position 1 to position 2, making a cut 0.13″ in depth.

Position 2

Position 1

0.1300

3.0000

0.7500

G02 and G03—Circular Motion

The G02 and G03 codes permit the cutter to travel along a circular path in a specific direction with a programmed feed rate. Before completing the command, the tool must be at the starting point of the circular move. G02 is a clockwise rotation and G03 is a counterclockwise rotation. The *radius method* and the *IJK method* are the two ways of programming a circular move.

An example command for the radius method is G03 X1.5 Y3.0 R1.0 F5.0. See **Figure 8-4**. The G03 indicates a counterclockwise direction, the X1.5 and Y3.0 specify the end point of circular arc, the R1.0 is the radius of circular arc, and F5.0 is the feed rate.

With the IJK method, the *I* is the distance and direction from the start point of the arc to the center of the arc along the X axis. The *J* is the distance and direction from the start point of the arc to the center of the arc along the Y axis. The *K* is the distance and direction from the start point of the arc to the center of the arc along the Z axis.

An example of a 90° arc command using the IJK method is G02 X2.5 Y2.0 I-1.0 J-1.0 F6.0. See **Figure 8-5**. The G02 indicates a clockwise direction; the X2.5 and Y2.0 specify the end point of circular point, the I-1.0 is the distance and direction from arc start point to arc center along X axis, J-1.0 is the distance and direction from arc start point to arc center along Y axis, and the F6.0 is the feed rate in inches per minute.

An example of a partial arc command using the IJK method is G02 X4.6562 Y1.2617 I-1.125 J-2.385. In **Figure 8-6**, this arc is shown as the path from N2 to N3. The G02 indicates a clockwise direction, the X4.6562 and Y1.2617 specify the end point of circular path, the I-1.125 is the distance and direction

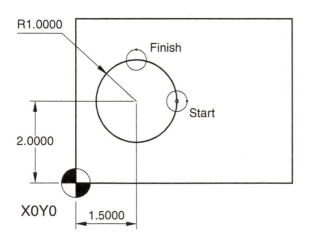

Figure 8-4. This is an example of the G03 command, which is used to cut a radius (curve) in a counterclockwise direction. The radius method of programming is shown.

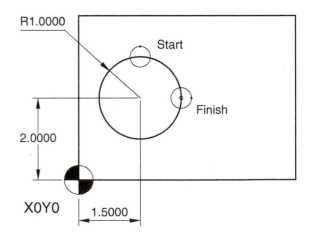

Figure 8-5. This is an example of the G02 command, which is used to cut a radius (curve) in a clockwise direction. The IJK method is shown.

from arc start point to arc center along X axis, and J-2.385 is the distance and direction from arc start point to arc center along Y axis. The following program example is for the partial arc move illustrated in Figure 8-6:

N1 G00X3.25Y4.25M03

N2 G01Y2.6651F10.

N3 G02X4.6562Y1.2617I-1.125J-2.385

N4 G01X5.75

Some controllers have the ability to machine a full circle. When performing this operation, the I-, J-, and K-words must be used. The following program example is for the full circle move illustrated in **Figure 8-7.**

N30 G00X0Y1.75

N35 G01Z-.125 F8.0

N40 G02J-1.75

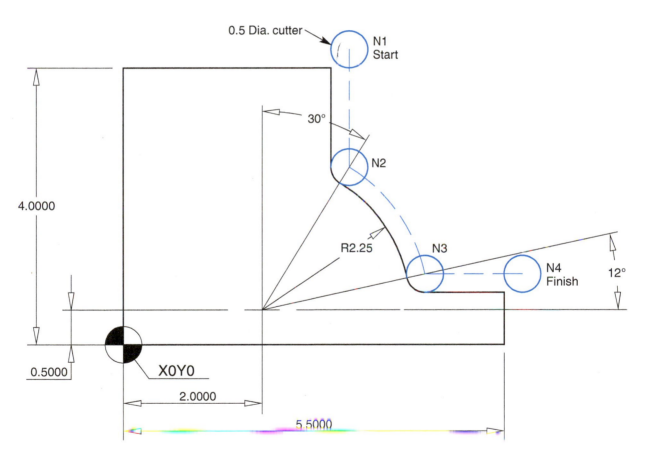

Figure 8-6. Here is an example of two straight cuts and a radius cut taken on a workpiece. The IJK method was used to program the cut along the arc from N2 to N3.

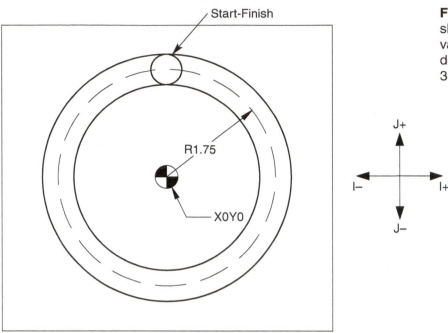

Figure 8-7. This illustration shows the directions of I and J values and a partial program depicting the machining of a 360° tool path.

G02 and G03—Helical Motion

Helical motion is primarily applied to thread milling. ***Thread milling*** is used on external and internal large threads, usually using a thread milling cutter. The cutter looks like a hog milling cutter and a tap combined. One machining pass is performed to form the thread. X and Y circular motion forms the thread diameter while Z axis motion forms the pitch of the thread. The machining process involves arcing the cutter into the workpiece to the proper depth (O.D. of thread), performing the helical movement to the required length, and finally exiting the workpiece with an out-arc movement.

Figure 8-8 shows the machining of a 2″ internal thread using a thread milling cutter. The program for the example follows:

O0007
N10G54G90S400M03
N20G00X1.5Y1.75   *Rapid move to Position 1*
N30G43H01Z.050   *Tool length compensation, rapid to .050″ above workpiece*
N40G42D11X.75F2.75   *Cutter radius compensation feed to Position 2*
N50G02X1.5Y2.5R.75   *Arc move to Position 3*
N60G02Y.5Z-.075R1.0   *Helical move to Position 4*
N70G02Y2.5Z-.200R1.0   *Helical move to Position 5*
N80G02Y.5Z-.375R1.0   *Helical move to Position 6*
N90G02Y2.5Z-.400R1.0   *Helical move to Position 7*

N100G02Y.5Z-.525R1.0	*Helical move to Position 8*
N110G02Y2.5Z-.650R1.0	*Helical move to Position 9*
N120G02Y.5Z-.775R1.0	*Helical move to Position 10*
N130G02X.75Y1.25R.75	*Arc move to Position 11*
N140G01G40X1.5	*Cancel radius compensation to Position 12*
N150G91G28Z0	*Retract spindle to home position*
N160G28X0Y0	*Return to home position in X and Y*
N170M3	*End of program*

G04—Dwell Command

The G04 *dwell command* causes axis motion to pause for a specific amount of time. This command is frequently applied in situations where tool pressure needs to be relieved. This is the case in spotfacing or when plunge cutting prior to machining a pocket. The following program example uses the dwell command:

N80G01Z-.100F2.5	*Plunge cut*
N90G04P750	*Pauses Z axis for .75 second*

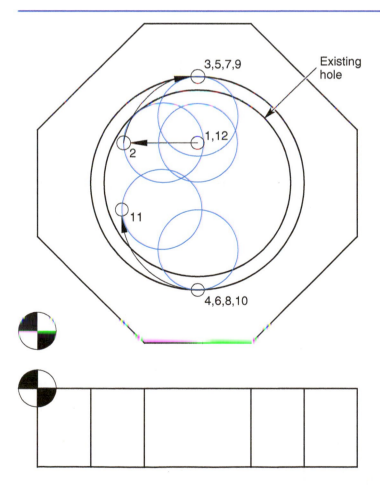

Figure 8-8. Helical motion can be used to cut a large-diameter hole without using an expensive single-point tool.

The P-word, used with the G04 word, specifies dwell time. The decimal point is not used with the P-word. Some examples of pause time include P500 (0.5 second), P1000 (1 second), and P1500 (1.5 seconds).

G09—Exact Stop Check

When a series of moves is made, the control automatically rounds corners between motions. The amount of rounding is almost unnoticed and is based on the feed rate and various parameters that are set. To make the control actually stop between commands, and thereby omit rounding, the G09 code is used.

```
N100G01G09X4.0F10
N110G01Y2.0
```

The tool travel stops after block N100, then resumes movement.

G10—Tool Offset Input by Program Command

The G10 command is used on machines that run programs with many tools. The offsets for these tools can be saved and retrieved from an outside source such as a computer. It is used with jobs that are run frequently, have many tools, and have tooling that remains on the machine.

G17, G18, and G19—Plane Selection Commands

Nearly all jobs are accomplished in the G17 (X,Y) plane. This code is initialized and therefore is not usually given in a program. The G18 (XZ) and G19 (YZ) commands are seldom used.

G20 and G21—Inch and Metric Modes

The G20 code indicates the job is done in inch mode, and the G21 code indicates the job is done in metric mode. The majority of jobs are performed in the inch system—all values such as zero location, tool lengths, and offsets must be stated in inches. G20 is typically the default status on machines.

G22 and G23—Stored Stroke Limit Commands

The G22 code sets up a zone (box or cube) around a machine device to prevent travel of a tool into that zone. The values of X, Y, Z, and I, J, K set the boundaries of the zone, **Figure 8-9**. An alarm sounds if the tool tries to enter the zone through an automatic operation or a manual movement. This code is not often used because it does not account for tool compensation. Therefore, a G22 would have to be set for each tool. The G23 code cancels the zone. The example N0010G22X-12Y-9Z-10I-16J-15K-14 sets a safety zone 4" long in X, 6" wide in Y, and 4" high in Z:

I16 to X12 is 4"
J15 to Y9 is 6"
K14 to Z10 is 4"
An N0010G23 command cancels stroke limit.

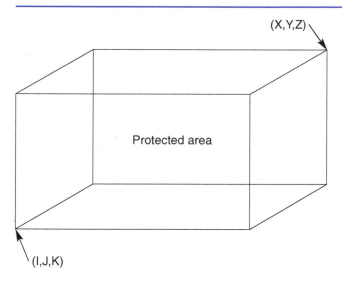

Figure 8-9. An example of a program-protected area defined by using X, Y, Z and I, J, K values.

G28—Zero Return Command

The G28 code is a two-part command that ultimately sends the machine to machine zero return, or *home*. The first part of the command designates which axes will be sent to an intermediate position. The second part of the command sends the machine from the intermediate position to machine home. Any axes that the operator wants to send home must be listed in the command. If an axis is not listed, it will not return home. The purpose of the intermediate position is to avoid any obstructions on the path to machine home. When using the G28 code, the G91 code (incremental mode) is used. Following are several examples of using the G28 code.

Example 1:

N0200G91G28X0Y0Z0   *Incrementally move nothing in all three axes but go directly to home in all three axes*

Example 2:

N0250G91G28Z0   *Incrementally move nothing in Z, return Z axis to home*

Example 3:

N0310G91G28X0Y0Z6.0   *Incrementally move nothing in X and Y, move up 6 inches in Z and send X and Y axes to home position*

G30—Return to Second Reference Position

The G30 code indicates another reference position made available by setting machine parameters. This command is sometimes used as a starting position for a program because it can be set closer than machine home. This command is not often used, however. The example N0200G91G30X0Y0X0 incrementally moves nothing in X, Y, and Z, then proceeds to the second reference point.

G40, G41, and G42—Tool Radius Compensation Commands

These codes adjust the position of the tool based on the radius of the tool. Tools may be worn to a smaller diameter. Use these codes to compensate for the difference.

When cutter compensation is not used, the tool path is calculated using the radius value of the cutter of the specified tool. If a .5″ end mill is needed to cut the profile on a part, the tool path will be .25″ away from the finished surface. If the diameter of the tool has been worn .020″, the finished part will be .010″ larger than intended. Using cutter compensation, the machinist measures the tool and enters this exact dimension. The tool path will be adjusted to match the radius of the worn tool, producing a part to the correct dimension. Other advantages to using cutter compensation include:

- Tool pressure, resulting from heavy cuts or dulling cutters, may cause tool deflection. Tool deflection causes oversize parts. With cutter compensation, if the part is out of tolerance, the situation can be remedied by changing the diameter value in the controller to change the amount of offset, thereby bringing the part within tolerance.

- Several cutting passes are easy to take when using cutter compensation. A dimension larger than the tool is entered for the rough cuts. The tool cuts farther away from the intended finish size. After each pass, the dimension is reduced until the exact diameter of the tool is entered. This dimension will cause the tool to cut the part to its final size.

Three codes apply to the use of cutter compensation. They are:

- **G40.** Cancels radius compensation

- **G41.** Cutter compensation left

- **G42.** Cutter compensation right

To determine whether to use the G41 or G42 code, the programmer should view the cutter from behind as it moves away. If the cutter is to the left of the surface being cut, it requires a G41 command; if the cutter is to the right of the surface being cut, it should be a G42 command. See **Figure 8-10**.

Invoking cutter compensation

After selecting whether G41 or G42 is to be used for cutter compensation, additional steps must be taken before invoking cutter compensation. The programmer should position the cutter as close to the workpiece as possible, but with a distance equal to slightly more than the radius of the cutter to allow for the offset move to take place before actual cutting occurs. To determine this distance, the programmer must have in mind what size cutter will be used. A right angle approach to the work should be made to allow the offset move to take place. Another step that must be taken to invoke cutter compensation is to command a D-word in the program that tells where the radius value of the tool is to be stored.

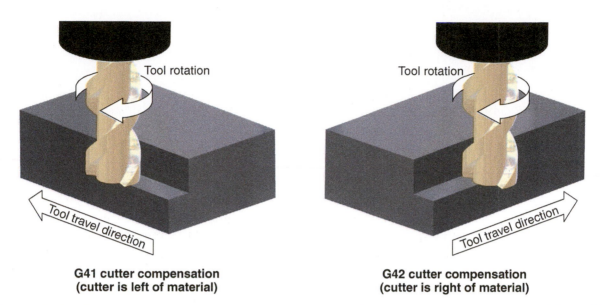

G41 cutter compensation
(cutter is left of material)

G42 cutter compensation
(cutter is right of material)

Figure 8-10. When you are looking behind the tool, if the tool is left of the material use G41, if the tool is right of the material use G42.

Remember, the actual surfaces of the workpiece are programmed; the cutter path runs parallel to these coordinates at a distance equal to the radius value in the stored D register. If a .5″ cutter is used to machine the part, the D value would be .25″. D offset number for cutter radius values should be different than tool length offsets. For example, if a tool length offset uses 2, then the radius value should use 32. This provides an organized method of dealing with various offsets within a program. It is good practice to invoke an offset off the workpiece whenever possible so the tool is already offset before the cutting begins. This eliminates any plunging into the work. See **Figure 8-11**.

Following is a simple program showing the format of how cutter compensation is programmed. The example shown will only refer to external machining.

O2001

N10G17G20G80G40G49	*Default block*
N20G90G54S1000M3T1M6	*Absolute system, fixture offset, spindle speed, CW rotation, tool change to tool 1*
N30G00X–.75Y–.5	*Rapid to position 1*
N40G90G43H01Z.2	*Tool length compensation, rapid above part*
N50M08	*Turn on coolant*
N60G01Z–.125F4.0	*Feed to depth*
N70G42D31Y.250F5.0	*Begin cutter compensation, feed to position 2*
N80X4.5	*Feed to position 3*
N90G03Y3.25R1.5	*Circular move to position 4*
N100G01X.75	*Feed to position 5*
N110G03X.25Y2.75R.5	*Circular move to position 6*

N120G01Y.75	*Feed to position 7*
N130G03X.75Y.25R.5	*Circular move to position 8*
N140G00Z.2	*Rapid to clearance plane*
N150G40	*Cancel cutter compensation*
N160G91G28Z0G49	*Return to Z home position, cancel tool length compensation*
N170G28X0Y0	*Return to X&Y home*
N180M30	*End and rewind program*

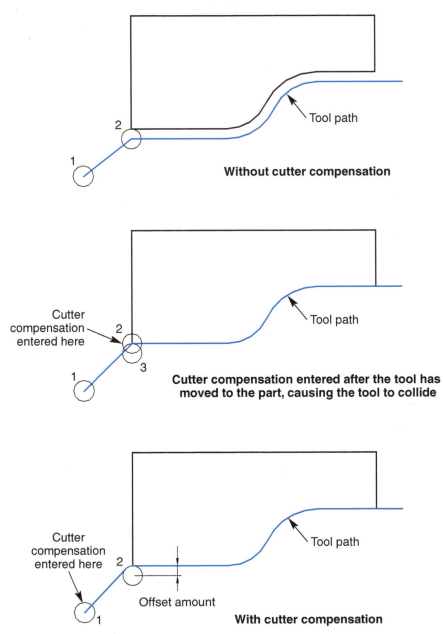

Without cutter compensation

Cutter compensation entered after the tool has moved to the part, causing the tool to collide

With cutter compensation

Figure 8-11. Without cutter compensatoin, the tool will create an accurate part only if the tool has not been worn to a smaller diameter. Cutter compensatoin should be entered before the tool moves to the part, or a collision will occur.

G43—Tool Length Compensation

The G43 code is used to enact tool length compensation for each tool. The command is used with an H-word, which holds the value of the offset. The H-word is usually the same as the tool number. The actual value of the tool offset is typically determined by the operator during the program setup and is entered into the offset values. For the example N0110G90G00G43H01Z.2, the tool is rapidly fed to a position of 0.2″ above part zero while the offset value H01 for that tool is applied to that movement.

Lengths vary from tool to tool. Therefore, the distance from the tip of the tool in a program will be different when measured to the Z program zero point (usually the top of workpiece). For example, if a counterbore hole is to be produced using a drill (6.342 length) and an end mill (3.755 length), the distance traveled in Z for the end mill will be greater because the end mill is shorter in length. See **Figure 8-12**.

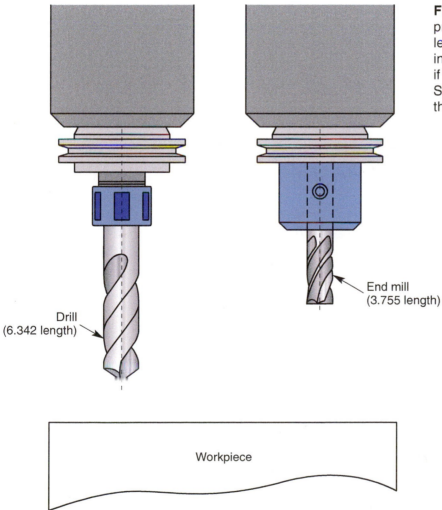

Figure 8-12. Tools used in a program will have different lengths. Offset values included in a program treat the tools as if they are the same length. Spindles are shown here in the fully retracted position.

End mill
(3.755 length)

Drill
(6.342 length)

Workpiece

Tool length compensation uses H offset values to theoretically make the tools equal in length. When traveling to a height of 1.0" above the workpiece, the end mill would have to travel an extra amount of 2.587" (drill length – end mill length) to reach the same position as the drill.

Tool offsets

Fanuc controls have at least 64 tool offsets that are stored in memory banks and contain tool length values. Values can be either length of the tool or the distance from the tip of the tool at spindle home position (zero return) down to part program zero. Tool offset (H-word) numbers should be the same as the tool station number to avoid errors and confusion in programming. For example, tool station 1 should be tool offset 1, tool station 2 should be tool offset 2, and so on.

Tool length compensation methods

Tool length compensation should be initialized in each tool's first axis move. This move is typically made to an R-plane position, usually 0.100" above workpiece zero. During this move, the tool offset value is invoked by the G43 code, which uses the H-word value to determine final Z movement. There are two methods of determining tool length compensation. The first method is shown in **Figure 8-13**:

1. Touch the spindle nose to the top of workpiece and record the Z value, **Figure 8-13A**. Input this value is to the fixture offset Z value memory bank.

2. Next, install a tool in the spindle and touch its tip to the top of the workpiece (Z0). See **Figure 8-13B**. Record the Z value for the tool touch and subtract that value from the spindle nose touch value. Repeat for each tool to be used. The result is tool length values that are input to the tool offset memory banks and used as H values.

For example, if the Z readout for spindle nose touch off is –13.650" (input into G54 Z offset) and the Z readout for drill touch is –10.755", the tool length is calculated by subtracting 10.755" from 13.650", or 2.895". The 2.895" is input as an H offset value into Z tool offset file.

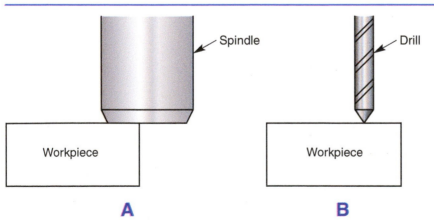

Workpiece

A

Drill

Workpiece

B

Figure 8-13. Both the spindle face and the tip of the drill are touched to the workpiece when setting tool length compensation.

Spindle

Caution

This method requires that offset values be + (positive), otherwise a crash would result.

The second method of figuring tool length compensation uses the distance from the tip of the tool to the top of the part (part zero). Each tool must be at the zero return (home) position before the tool touches the top of the part (Z zero). Each value will be a negative (–) value, therefore, each offset value will be negative as well. If lengths vary from tool to tool, the distance from the tool tip of each tool in a program will be different when measured to the Z program zero point (usually top of workpiece). For example, if a counterbored hole is produced using a drill and an end mill, the distance moved in Z for the end mill is greater because the end mill is shorter in length. Tool length compensation uses H offset values to theoretically make the tools equal in length. In **Figure 8-14**, when traveling to a distance 1″ above the workpiece, the end mill would have to travel 2.587″ (6.342 – 3.755) further to reach the same position as the drill.

G44—Tool Length Compensation

The G44 code is similar to the G43 code except the + or – directions of the offset are reversed. This command is not as popular as the G43 command.

G49—Tool Length Compensation Cancel

The G49 code cancels the tool length offset for that tool. It is used when the spindle is sent to the Z home position. This command is not always used, since the G43 or G44 code for that tool is replaced by the G43 or G44 for the next tool.

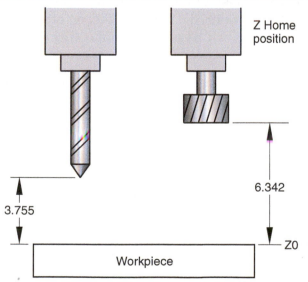

Figure 8-14. Tool length compensation uses different H values for a drill and a milling cutter to treat them equal in length. Tools shown are in a fully retracted (home) position.

G50 and G51—Scaling

Scaling multiplies the program's X, Y, and Z values by a scale factor amount, thereby reducing or enlarging the part. Scaling is used primarily in the die and mold industries. It is used on a family of parts, such as parts that are identical in shape but different in size. The parts must be three-dimensional in nature. Only axis motions are changed in a program. The codes used with scaling include G50, which turns on the scaling function; I, the scale center in X; J, the scale center in Y; K, the scale center in Z, and P, the scale factor. The G51 command cancels scaling.

G54 through G59—Fixture Offsets

Fixture offsets, or work coordinate settings, assign program zero to a workpiece. The distances from the machine's zero return position to the program zero point are the values entered in the fixture offset page of the Fanuc control. See **Figure 8-15**. The offset values will be minus values.

Any of these codes can be used to set part zero from machine home. The program runs from part zero. Usually, G54 is used to designate part zero on a single workpiece. However, when multiple parts are run from a program, several codes are used to set part zero for each part.

The best method to use for setting part zero is to place the code that sets part zero at the beginning of the program. This technique allows the program to start at any point and still arrive at the proper coordinates. This is important to know if for some reason a program has to be picked up in the middle and run. Some jobs might require the use of multiple fixture offsets to machine several parts using one program. See **Figure 8-16**.

G68 and G69—Coordinate Rotation

The G68 code permits the programmer to describe the shape of the workpiece to be machined in a 90° and 180° position, then rotate the shape to the actual machining position. This command contains an R-word, which tells the control the angle at which the rotation should be made. Every motion following the G68 code will be rotated until a G69 (cancel rotation) command is given. An example coordinate rotation command is:

N075G68R40   *Rotates to a 40° angle*
N110G69   *Cancel rotation*

G98 and G99—Initial Plane and R-plane Commands

The *initial plane* is the last Z position of tool prior to the canned cycle command. The *R-plane* is the R value found in the canned cycle command. The initial plane and R-plane commands are used with canned cycle commands to clear clamps and obstructions. The G98 code retracts the tool to the initial plane. The G99 code retracts the tool to the R-plane.

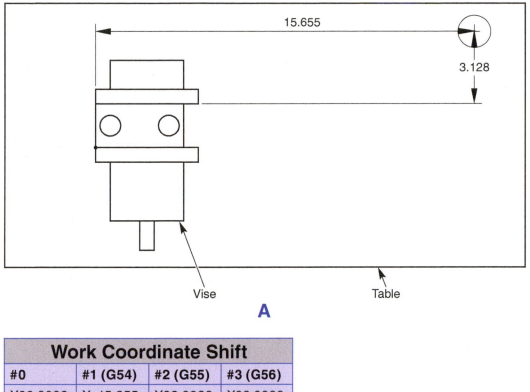

Work Coordinate Shift			
#0	#1 (G54)	#2 (G55)	#3 (G56)
X00.0000	X–15.655	X00.0000	X00.0000
Y00.000	Y–3.128	Y00.000	Y00.000
Z00.0000	Z00.0000	Z00.0000	Z00.0000

B

Figure 8-15. Fixture offsets. A—The location of a vise in relation to machine home position determines its fixture offset value. B—An example of a fixture offset page on a Fanuc machine controller.

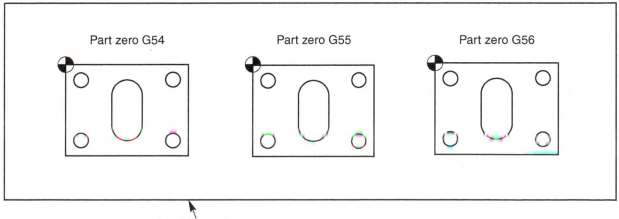

Figure 8-16. The location of three fixtures on a machine table, relative to machine home, are defined using three G-codes (G54, G55, and G56).

M-Codes

The *M-codes* control miscellaneous functions in a program. They can be modal or nonmodal. There are many M-codes for various machines; the most frequently used codes are discussed here. These include the following:

- **M00.** This is the *program stop command*. The program halts until the cycle start button is pressed. This code is used to reclamp a workpiece, apply cutting compound, reposition a part, or remove chips. The M00 code may be found within a program more than once to allow the operator to perform a task.

- **M01.** The *optional stop command*, which permits an operator to stop the program if desired. This command is usually used after each tool finishes cutting. This allows the operator to inspect the operation, especially during setup. A switch must be turned on to recognize the code.

- **M02.** The *end of program command* is used when a program runs with punched tape. This is the last command in the program and it rewinds the program to the beginning.

- **M03.** The *spindle on clockwise command* turns on machine spindle in clockwise rotation.

- **M04.** The *spindle on counterclockwise command* turns on machine spindle in counterclockwise rotation.

- **M05.** The *spindle off command*.

- **M06.** The *tool change command*. M06 performs an automatic tool change. This command will remove a tool from the spindle and insert a new tool when used with a T-word. Example: T2 M6.

- **M08.** The *turn coolant on command*.

- **M09.** The *turn coolant off command*.

- **M30.** This is the *end of the program command*. It stops all machine operations and rewinds the program to the start position.

- **M98.** This is the *subprogram jump to command*. It tells control to jump to a subprogram. Used with P-word (the subprogram number) and an L-word, which states how many times to repeat the subprogram. Subprograms contain information that would have to be repeated several times in the main program. The use of subprograms reduces the total length of a program. Subprograms will be covered in detail in *Chapter 16*.

- **M99.** The *subprogram jump back command*. This is the command at the end of a subprogram that returns control to the next command in the main program.

S-, T-, F-, H-, and D-Words

Word addresses beginning with the letters *S, T, F, H,* and *D* provide important information on topics such as spindle speed, feed rate, and offsets for compensation. These words were covered earlier in this chapter in a listing of word addresses, but are repeated here for review.

- **S-word:** Spindle speed

  Format: S*xxxx*

  Examples: S1000 (1000 rpm)

  S650 (650 rpm)

- **T-word:** Tool code (designates location of tool)

  Format: T*xx*

  Examples: T05 (Tool 5)

  T1 (Tool 1)

- **F-word:** Feed rate in inches per minute (ipm)

  Format: F*xx*

  Examples: F12 (12 ipm feed)

  F7.5 (7.5 ipm feed)

- **H-word:** Offset number for tool length compensation

  Format: H*xx*

  Examples: H1 (Tool length offset 1)

  H08 (Tool length offset 8)

- **D-word:** Offset number for cutter radius compensation

  Format: D*xx*

  Examples: D6 (Radius compensation offset 6)

  D07 (Radius compensation offset 7)

Summary

Alpha characters and numbers form words that are combined to form a block. These blocks give a command to the computer to perform a function.

An address is an alphabetic character that has many meanings throughout a program, such as preparatory, coordinate position, and miscellaneous functions. G-codes are called preparatory functions and place the machine in various operating modes. These codes may be modal or nonmodal. Several are initialized when the machine power is turned on.

Tool length compensation is used in a program to make tools of various lengths appear to the control as one length. Compensation ensures that if all tools were sent to the same specific height, the end of each tool would be at that position. Fixture offsets are used to tell the machine where part zero is located on the workpiece.

M-codes are like on/off switches that control various functions such as spindle on/off and coolant on/off. M-codes can be modal or nonmodal.

Chapter Review

Answer the following questions. Write your answers on a separate sheet of paper.

1. Provide a brief description of the following alphabetic characters indicated in a program.
   a. O
   b. N
   c. G
   d. X
   e. U
   f. Y
   g. W
   h. Z
   i. R
   j. Q
   k. P
   l. L
   m. F
   n. S
   o. T
   p. M
   q. D
   r. H
2. List three advantages of using cutter compensation.
3. What code allows cutter compensation when the tool is traveling away from you and is right of the material?
4. What code cancels cutter compensation?
5. List the G-codes needed for the various modes given below.
   a. Rapid traverse
   b. Straight motion
   c. CW circular motion
   d. CCW circular motion
   e. Dwell
   f. Inch mode
   g. XZ plane
   h. Zero return
   i. Tool length compensation
   j. Fixture offset
6. List the M-codes needed to perform the following functions.
   a. Program stop
   b. Optional stop
   c. Spindle on clockwise
   d. Spindle on counterclockwise
   e. Spindle off
   f. Coolant on
   g. Coolant off
   h. Tool change

7. State the word needed to give each of the following commands to the machine controller.
   a.   Feed rate of 15 inches
   b.   Tool length offset for tool 8
   c.   Spindle speed of 1000 rpm
   d.   Cutter compensation offset for tool 6

Activities

1. Obtain various programs from different machine controllers and compare words as to their arrangement (order) and uses. Check for new codes not discussed in this chapter, and attempt to identify and discuss the purpose of these codes.

The Precision Machined Products Association is an international trade association representing the interests of the precision machined products industry. www.pmpa.org

Chapter 9
Program Format for Vertical Machining Centers

Objectives

Information in this chapter will enable you to:

- Write a machining center program without using canned cycles.
- Describe the effect of various codes in a block of information.

Program Format

There can be several variations of program formats. Companies use the format that works best for their particular purposes. The format used in this textbook is one that works with a carousel-type tool changer. The tool changer expects the program to be started from a home position using fixture offsets. Defaults of a machine may make it possible to eliminate some codes found in this format. In addition, some codes may be eliminated to shorten this format if desired. This format can be customized for the user's particular needs. The format is divided into four sections: the program start, tool ending, tool start, and program end. The following are examples of each of these sections:

Program Start

O0010	*Program number*
N10G17G20G80G40G49	*Default block – line number, X and Y plane, inch input, cancel canned cycle, cancel radius compensation, cancel tool length compensation*
N20T2M6	*Tool change*
N30G90G54S500M03	*Absolute system, fixture offset, spindle speed, clockwise rotation*
N40G00X1.0Y1.0	*Rapid move to first position*
N50M01	*Optional stop*
N60G43H02Z.2	*Tool length compensation, offset 2, rapid plane*
N70M08	*Turn coolant on*

Tool Ending

N90G00Z.2M09	*Rapid to R-plane, turn off coolant*
N100G91G28Z0G49	*Rapid to reference point return, cancel tool length compensation*

Tool Start

N190G17G20G80G40G49	*Default block*
N200T3M6	*Tool change to tool 3*
N300G90G54S600M03	*Absolute system, fixture offset, spindle speed, CW rotation*
N310G00X2.Y2	*Rapid to position*
N320M01	*Optional stop*
N330G43H03Z.2	*Tool length compensation, offset 3, rapid plane*
N340M08	*Turn coolant on*

Program End

N400M9	*Turn coolant off*
N410G91G28Z0G49	*Return to Z home position, cancel tool length compensation*
N420G91G28X0Y0	*Return to X and Y home position*
N430M30	*End and rewind program*

The format being used allows the operator to start the program with any tool simply by selecting the default line preceding the tool desired. Actually, each tool in the program is a program by itself.

Simple Drilling Program—Absolute System

Refer to **Figures 9-1** and **9-2** as you study the sample program. Information regarding setup instructions, tooling, or other data can be provided to the CNC operator by inserting that information within the body of the program. This information is not acted on by the machine controller. Information is furnished to the operator by putting it between parentheses.

O0001 (Cover Plate)	*Project number and name*
N10G17G20G80G40G49	*Default block*
N20T1M6 (No. 3 centerdrill)	*Tool change – tool 1*
N30G90G54S3000M03	*Absolute system, fixture offset, spindle speed, CW rotation*
N40G00X1.0Y1.0 (Position 1)	*Rapid to position 1*
N50M01	*Optional stop*
N60G43H01Z.2	*Tool length comp, rapid to clearance plane*
N70M08	*Coolant on*
N80G01Z–.250F2.0	*Feed to Z–.25 F2.0 = 2 ipm*
N90G00Z.2	*Rapid to clearance plane*

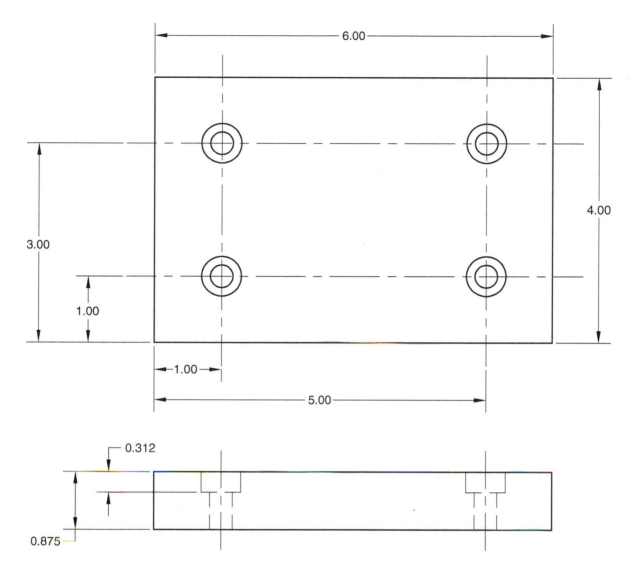

Figure 9-1. This cover plate, used to seal an opening, has examples of drilling and counterboring operations.

Order	Tool	Tool station	rpm	Feed
1	centerdrill	1	3000	2.0
2	3/8" drill	2	1200	3.5
3	0.5" end mill	3	800	4.0

Figure 9-2. Tools used, and their order of use, are listed on a tool sheet. Also listed are the tool locations, and the speeds and feeds used.

N100G00Y3.0 (Position 2)	*Rapid to position 2*
N110G01Z–.250F2.0	*Feed to Z–.25 F2.0 = 2 ipm*
N120G00Z.2	*Rapid to clearance plane*
N130G00X5.0 (Position 3)	*Rapid to position 3*
N140G01Z–.250F2.0	*Feed to Z–.25 F2.0 = 2 ipm*
N150G00Z.2	*Rapid to clearance plane*
N160G00Y1.0 (Position 4)	*Rapid to position 4*
N170G01Z–.250F2.0	*Feed to Z–.25 F2.0 = 2 ipm*
N180G00Z.2M09	*Rapid to clearance plane, coolant off*
N190G91G28Z0G49	*Rapid to reference point return, cancel tool length compensation*
N200G17G20G80G40G49	*Default block*
N205T2M6	*Tool change – tool 2*
N210G90G54S1200M03 (3/8 drill)	*Absolute system, fixture offset*
	Spindle speed, CW rotation
N220G00X1.0Y1.0 (Position 1)	*Rapid to position 1*
N230M01	*Optional stop*
N240G43H02Z.2	*Tool length compensation, rapid to clearance plane*
N250M08	*Turn coolant on*
N260G01Z–1.1F3.5	*Feed thru part, F3.5 = 3.5 ipm*
N270G00Z.2	*Rapid to clearance plane*
N280G00Y3.0 (Position 2)	*Rapid to position 2*
N290G01Z–1.1F3.5	*Feed thru part, F3.5 = 3.5 ipm*
N300G00Z.2	*Rapid to clearance plane*
N310G00X5 (Position 3)	*Rapid to position 3*
N320G01Z–1.1F3.5	*Feed thru part, F3.5 = 3.5 ipm*
N330G00Z.2	*Rapid to clearance plane*
N340G00Y1.0 (Position 4)	*Rapid to position 4*
N350G01Z–1.1F3.5	*Feed thru part, F3.5 = 3.5 ipm*
N360G00Z.2M09	*Rapid to clearance plane, coolant off*
N370G91G28Z0G49	*Rapid to reference point return, cancel tool length compensation*
N380G17G20G80G40G49	*Default block*
N385T3M6 (.5 end mill)	*Tool change – tool 3*
N390G90G54S800M03	*Absolute system, fixture offset, spindle speed, CW rotation*
N400G00X1.0Y1.0 (Position 1)	*Rapid to position 1*
N410M01	*Optional stop*
N420G43H03Z.2	*Tool length comp, rapid to clearance plane*
N430M08	*Turn coolant on*
N440G01Z–.312F4.0	*Feed to .312 depth, F4.0 = 4 ipm*
N450G04P500	*Pause Z axis for .5 seconds*
N460G00Z.2	*Rapid to clearance plane*
N470G00Y3.0 (Position 2)	*Rapid to position 2*

N480G01Z–.312F4.0	*Feed to .312 depth, F4.0= 4 ipm*
N490G04P500	*Pause Z axis for .5 seconds*
N500G00Z.2	*Rapid to clearance plane*
N510G00X5 (Position 3)	*Rapid to position 3*
N520G01Z–.312F4.0	*Feed to .312 depth, F4.0=4 ipm*
N530G04P500	*Pause Z axis for .5 seconds*
N540G00Z.2	*Rapid to clearance plane*
N550G00Y1.0 (Position 4)	*Rapid to position 4*
N560G01Z–.312F4.0	*Feed to .312 depth, F4.0=4 ipm*
N570G04P500	*Pause Z axis for .5 seconds*
N580M9	*Turn coolant off*
N590G91G28Z0G49	*Return to home position, cancel tool length compensation*
N600G91X0Y0	*Return to X and Y home positions*
N610M30	*End and rewind program*

NOTE

Various codes were repeated, but were not necessary since they are modal. However, repeating the codes helps us understand what is happening and aids in remembering the function of these codes.

Simple Drilling Program—Incremental System

This is a simple one-tool program showing the use of the incremental system of positioning. See **Figure 9-3**. This will be the only time the incremental system will be used in programming. It is seldom used, except in subprogramming, because an error in position values will carry throughout the entire program. Nearly all programming done uses the absolute system of positioning.

O0002 (Router Bit Plate)	*Project number and name*
N10G17G20G80G40G49	*Default block*
N15T1M6 (1/4 drill)	*Tool change – tool 1*
N20G90G54S1600M03	*Absolute system, fixture offset, spindle speed, CW rotation*
N30G91G00X1.0Y1.0 (Position 1)	*Incremental, rapid to position 1*
N40M01	*Optional stop*
N50G90G43H01Z.2	*Absolute, tool length compensation, rapid to clearance plane*
N60M08	*Turn on coolant*
N70G01Z–.5F3.5	*Feed to .5 depth, F3.5 = 3.5 ipm*
N80G00Z.2	*Rapid to clearance plane*
N90G91G00X1.0 (Position 2)	*Incremental, rapid to position 2*
N100G90G01Z–.5F3.5	*Absolute, feed to .5 depth, F3.5 = 3.5 ipm*

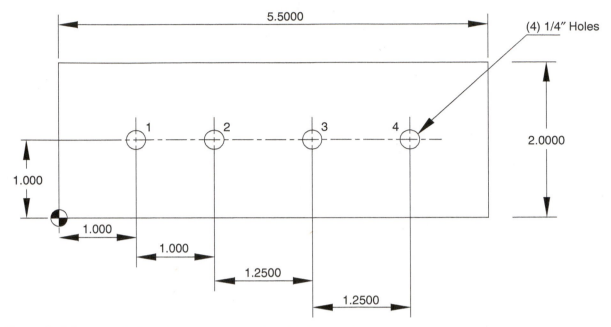

Figure 9-3. This router bit plate is used to store 1/4″ router bits.

N110G00Z.2	*Rapid to clearance plane*
N120G91G00X1.25 (Position 3)	*Incremental, rapid to position 3*
N130G90G01Z–.5F3.5	*Absolute, feed to .5 depth F3.5 = 3.5 ipm*
N140G00Z.2	*Rapid to clearance plane*
N150G91G00X1.25 (Position 4)	*Incremental, rapid to position 4*
N160G90G01Z–.5F3.5	*Absolute, feed to .5 depth F3.5 = 3.5 ipm*
N170M9	*Turn coolant off*
N180G91G28Z0G49	*Return to Z home position, cancel tool length compensation*
N190G91G28X0Y0	*Return to X and Y home position*
N200M30	*End and rewind program*

Simple Milling—Example 1

This milling example contains additional information given to the operator. See **Figure 9-4**.

O0003 (Mill Example 1)	
N10 (X0Y0 is located in the lower-left hand corner of the workpiece)	
N20 (Z0 is located on the top of the workpiece)	
N30 (Tool 1 is a .5″ end mill)	
N40G17G20G80G40G49	*Default block*
N45T1M6	*Tool change*

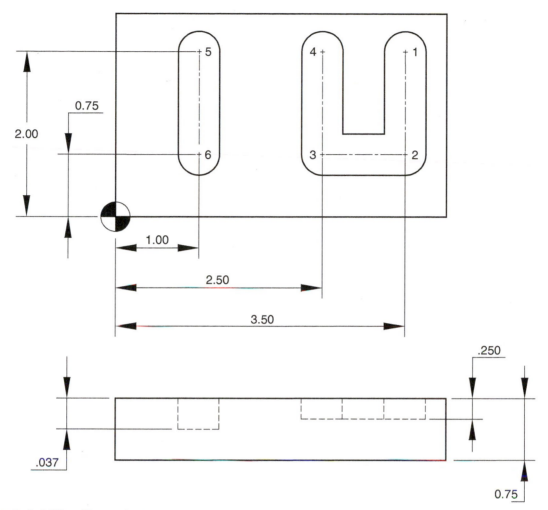

Figure 9-4. Milling Example 1.

N50G90G54S1600M03	*Absolute system, fixture offset, spindle speed, clockwise rotation*
N60G00X3.5Y2.0	*Rapid move to position 1*
N70M01	*Optional stop*
N80G43H01Z.2	*Tool length compensation, offset 1, rapid plane*
N90M08	*Turn coolant on*
N100G01Z–.25F3.5	*Plunge feed to .25 depth, 3.5 ipm feed rate*
N110X3.5Y.75	*Feed to position 2*
N120X2.5	*Feed to position 3*
N130X2.5Y2.0	*Feed to position 4*
N140G00Z.2	*Rapid to clearance plane*
N150X1.0Y2.0	*Rapid to position 5*
N160G01Z–.37	*Plunge feed to .37 depth, 3.5 ipm feed rate*
N170X1.0Y.75	*Feed to position 6*
N180G00Z.2M09	*Rapid to clearance plane, turn off coolant*

N190G91G28Z0G49	*Rapid to reference point return, cancel tool length compensation*
N200G91G28X0Y0	*Return to X and Y home position*
N210M30	*End and rewind program*

Simple Milling—Example 2

This example uses three tools and involves drilling and outside contouring with circular interpolation. See **Figures 9-5, 9-6**, and **9-7**. This program does not provide work holding information. Actual machining of the part requires two setups. The program would have to be altered to account for part fixturing. This information could be given within the program or on a setup sheet. The part requires pre-machining to remove excess material.

Order	Tool	Tool station	rpm	Feed
1	centerdrill	1	3000	2.0
2	1/2″ drill	2	2000	3.0
3	0.5″ end mill	3	1800	4.5

Figure 9-5. The tool sheet for Milling Example 2.

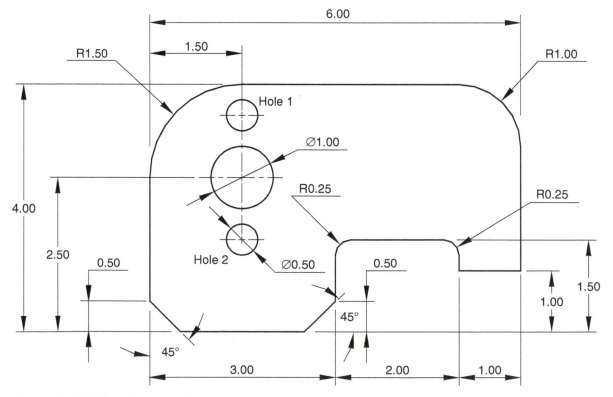

Figure 9-6. Milling Example 2.

O0004 (Mill Example 2)
N10 (X0Y0 is located at an existing 1″ reamed hole)
N20 (Z0 is located on the top of the part)
N30 (Tool 1 is a No. 3 centerdrill)
N40 (Tool 2 is a 1/2″ drill)
N50 (Tool 3 is a .500 end mill)

N60G17G20G80G40G49	*Default block*
N65T1M6 (No. 3 centerdrill)	*Spindle change – tool 1*
N70G90G54S3000M3	*Absolute speed, CW rotation, tool change, Fixture offset*
N80G00X0Y1.0 (Hole 1)	*Rapid to hole 1*
N90M01	*Optional stop*
N100G43H01Z.2	*Tool length compensation, offset 1, rapid plane*
N110M08	*Coolant on*
N120G01Z–.312F2.0	*Centerdrill hole, feed = 2 ipm*
N130G00Z.2	*Retract to rapid plane*
N140X0Y–1.0 (Hole 2)	*Rapid to hole 2*
N150M01	*Optional stop*
N160G43H02Z.2	*Tool length compensation, offset 2, rapid plane*

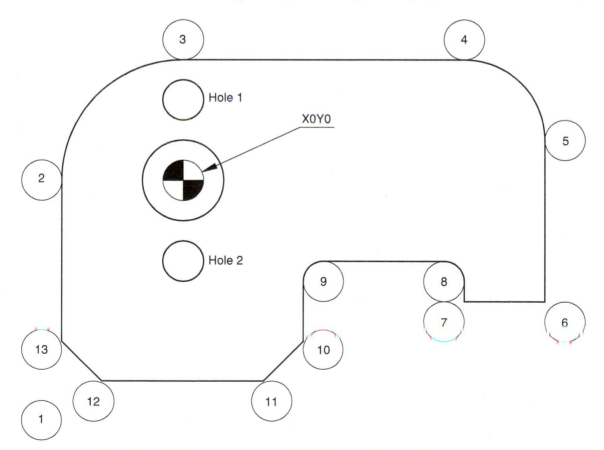

Figure 9-7. This print shows tool positions as referenced in Milling Example 2.

N170G01Z–.312F2.0	*Centerdrill hole, feed = 2 ipm*
N180G00Z.2M09	*Rapid to R-plane, coolant off*
N190G91G28Z0G49	*Rapid to reference point return, cancel tool length compensation*
N200G17G20G80G40G49	*Default block*
N205T2M6 (1/2" drill)	*Tool change – tool 2*
N210G90G54S1600M3	*Absolute, fixture offset, spindle speed, CW rotation*
N220G00X0Y1.0 (Hole 1)	*Rapid to hole 1*
N230M01	*Optional stop*
N240G43H02Z.2	*Tool length compensation, offset 2, rapid plane*
N250M08	*Coolant on*
N260G01Z–.650F3.5	*Drill thru part, feed = 3.5 ipm*
N270G00Z.2	*Retract to rapid plane*
N280X0Y–1.0	*Rapid to hole 2*
N290M01	*Optional stop*
N300G43H02Z.2	*Tool length compensation, offset 2, rapid plane*
N310G01Z–.650F3.5	*Drill thru part, feed = 3.5 ipm*
N320G00Z.2M09	*Retract to rapid plane, coolant off*
N340G91G28Z0G49	*Rapid to reference point return, cancel tool length compensation*
N350G17G20G80G40G49	*Default block*
N355T3M6 (.5 end mill)	*Tool change – tool 3*
N360G90G54S1800M03	*Absolute, fixture offset, spindle speed, CW rotation*
N370G00X–1.75Y–3.0	*Rapid to position 1*
N380M01	*Optional stop*
N390G43H03Z.2	*Tool length compensation, offset 3, rapid plane*
N400M08	*Coolant on*
N410G01Z–.500F3.5	*Feed to –.5" depth, feed = 3.5 ipm*
N420X–1.75Y0	*Linear feed to position 2*
N430G02X0Y1.75I1.75J0	*Circular arc to position 3*
N440G01X3.5Y1.75	*Linear feed to position 4*
N450G02X6.25Y.5I0J–1.25	*Circular arc to position 5*
N460G01X6.25Y–1.75	*Linear feed to position 6*
N470X3.25Y–1.75	*Linear feed to position 7*
N480X3.25Y–1.25	*Linear feed to position 8*
N490X1.75Y–1.25	*Linear feed to position 9*
N500X1.75Y–1.89	*Linear feed to position 10*
N510X1.110Y–2.75	*Linear feed to position 11*
N520X–.890Y–2.75	*Linear feed to position 12*
N530X–1.75Y–1.89	*Linear feed to position 13*
N540M09	*Coolant off*
N550G91G28Z0G49	*Return to Z home position, cancel tool length compensation*
N560G91G28X0Y0	*Return to X and Y home position*
N570M30	*End and rewind program*

Simple Milling—Example 3

This example uses four tools and involves operations such as drilling, slotting, pocketing, and circular profiling. This program will be shortened by using leading zero suppression and eliminating coordinate values that are not necessary. For example, if the last position of X or Y is the same, it will not be repeated. See **Figures 9-8**, **9-9**, and **9-10**. This workpiece would be held in a vise. The blank is 3.25" × 4.25".

Order	Tool	Tool station	rpm	Feed
1	centerdrill	1	3000	2.0
2	.375" drill	2	1600	3.5
3	.375" end mill	3	2000	5.0
4	.5" end mill	4	1800	4.0
5	0.75" end mill	5	1400	3.5

Figure 9-8. The tool sheet for Milling Example 3.

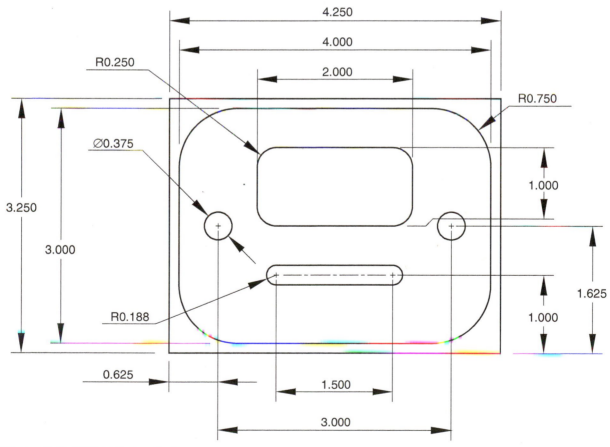

Figure 9-9. Milling Example 3.

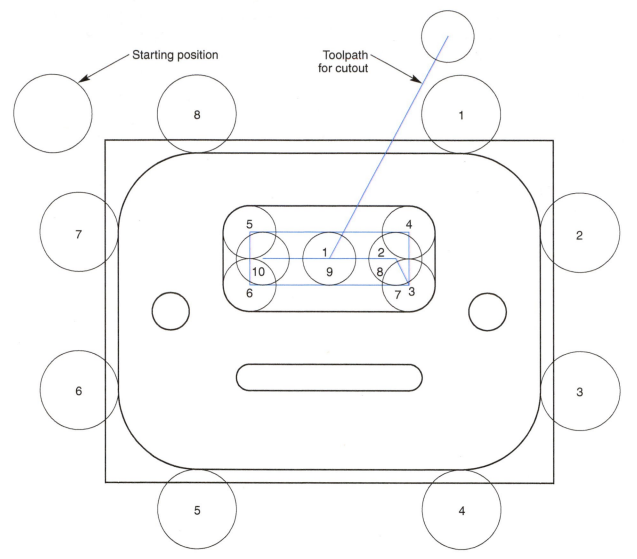

Figure 9-10. This print shows tool positions as referenced in Milling Example 3.

O4 (Mill Example 3)

N10 (X0Y0 is located in the lower-left corner of the workpiece)

N20 (Z0 is located on the top of the workpiece)

N30 (Tool 1 is a No. 3 centerdrill)

N40 (Tool 2 is a 3/8″ drill)

N50 (Tool 3 is a .375″ HSS end mill)

N60 (Tool 4 is a .5″ HSS end mill)

N70 (Tool 5 is a .75″ HSS end mill)

(Centerdrilling)

N80G17G20G80G40G49   *Default block*

N85T1M6 (No. 3 centerdrill)   *Tool change–tool 1*

N90G90G54S3000M3	*Absolute, fixture offset, spindle speed, CW rotation*
N100G0X.625Y1.625 (Left hole)	*Rapid to left hole*
N110M1	*Optional stop*
N120G43H1Z.2	*Tool length compensation, offset 1, rapid plane*
N130M8	*Turn coolant on*
N140G1Z−.312F2	*Drill to −.312 depth, feed = 2 ipm*
N150G0Z.2	*Retract to rapid plane*
N160X3.625	*Rapid to right hole*
N170G1Z−.312F2	*Drill to −.312 depth, feed = 2 ipm*
N180G0Z.2M9	*Rapid to R-plane, coolant off*
N190G91G28Z0G49	*Rapid to reference point return, cancel tool length compensation*
(Drilling)	
N200G17G20G80G40G49	*Default block*
N205T2M6 (.375 drill)	*Tool change–tool 2*
N210G90G54S1600M3	*Absolute, fixture offset, spindle speed, CW rotation*
N220G0X.625Y1.625 (Left hole)	*Rapid to left hole*
N230M1	*Optional stop*
N240G43H2Z.2	*Tool length compensation, offset 2, rapid plane*
N250M8	*Turn coolant on*
N260G1Z−.650F3.5	*Drill thru*
N270G0Z.2	*Retract to rapid plane*
N280X3.625 (Right hole)	*Rapid to right hole*
N290G1Z−.650F3.5	*Drill thru*
N300G0Z.2M9	*Rapid to R-plane, coolant off*
N310G91G28Z0G49	*Rapid to reference point return, cancel tool length compensation*
(Cutting Slot)	
N320G17G20G80G40G49	*Default block*
N325T3M6 (.375 end mill)	*Tool change–tool 3*
N330G90G54S2000M3	*Absolute, fixture offset, spindle speed, CW rotation*
N340G0X1.375Y1.125	*Rapid to slot start*
N350M1	*Optional stop*
N360G43H3Z.2	*Tool length compensation, offset 3, rapid plane*
N370M8	*Turn coolant on*
N380G1Z−.1875F3.5	*Plunge cut to −.1875 depth, feed = 3.5 ipm*
N390X2.875	*Linear feed–cut slot*
N400G0Z.2M9	*Rapid to R-plane, turn off coolant*
N410G91G28Z0G49	*Rapid to reference point return, cancel tool length*
(Cut Pocket)	
N420G17G20G80G40G49	*Default block*
N425T4M6 (.5 end mill)	*Tool change–tool 4*
N430G90G54S1800M3	*Absolute, fixture offset, spindle speed, CW rotation*

N440G0X2.125Y2.125	*Rapid to above position 1*
N450M1	*Optional stop*
N460G43H4Z.2	*Tool length compensation, offset 4, rapid plane*
N470M8	*Turn coolant on*
N480G1Z–.125F3.5	*Plunge cut to –.125 depth, feed = 3.5 ipm*
N490X2.75F5	*Cut to position 2*
N500X2.875Y1.875	*Cut to position 3*
N510Y2.375	*Cut to position 4*
N520X1.375	*Cut to position 5*
N530Y1.875	*Cut to position 6*
N540X2.875	*Cut to position 7*
N550X2.75Y2.125	*Cut to position 8*
N560X2.125	*Cut to position 9*
N570X1.5Y2.125	*Cut to position 10*
N580G0Z.2M9	*Retract to R-plane, turn off coolant*
N590G91G28Z0G49	*Rapid to reference point return, cancel tool length*
(Cut Outside Contour)	
N600G17G20G80G40G49	*Default block*
N605T5M6 (.75" end mill)	*Tool change – tool 5*
N610G90G54S1400M3	*Absolute, fixture offset, spindle speed, CW rotation*
N620G0X–.5Y3.5	*Rapid to starting position*
N630M1	*Optional stop*
N640G43H5Z.2	*Tool length compensation, offset 5, rapid plane*
N650M8	*Coolant on*
N660G1Z–.125F3.5	*Feed to –.125 depth, feed = 3.5 ipm*
N670X3.375	*Linear feed to position 1*
N680G2X4.5Y2.375J–1.125	*Circular feed to position 2*
N690G1Y.875	*Linear feed to position 3*
N700G2X3.375Y–.25I–1.125	*Circular feed to position 4*
N710G1X.875Y–.25	*Linear feed to position 5*
N720G2X–.25Y.875J1.125	*Circular feed to position 6*
N730G1Y2.375	*Linear feed to position 7*
N740G2X.875Y3.5	*Circular feed to position 8*
N750M9	*Turn coolant off*
N760G91G28Z0G49	*Return to Z home position, cancel tool length compensation*
N770G91G28X0Y0	*Return to X and Y home position*
N780 M30	*End and rewind program*

Summary

A programming format for a machining center includes a program start, tool ending, tool start, and program ending. Information for the operator regarding setup, tooling, fixturing, and other pertinent information can be provided by placing this information within parentheses in the body of the program.

Chapter Review

Answer the following questions. Write your answers on a separate sheet of paper.

1. How many blocks of information are used to accomplish a tool ending procedure?
2. Write the block information needed to send a tool to a position .100″ above a hole.
3. Write the block information that sends the Z axis home.
4. Write the block information used as the default block in a program.
5. Write the block information used to remove a tool from a hole.
6. Write the block information that makes a tool change for tool 7.

Activities

1. Complete the following programming activity.
   A. Determine the coordinate positions shown on the clamp block print. Place these coordinate values in the table. Note: Position the cutter for end mill approach and exiting 0.750" away from the end of the workpiece.
   B. Use the information from above to determine the codes, spindle speeds, and feed rates missing in the program provided.

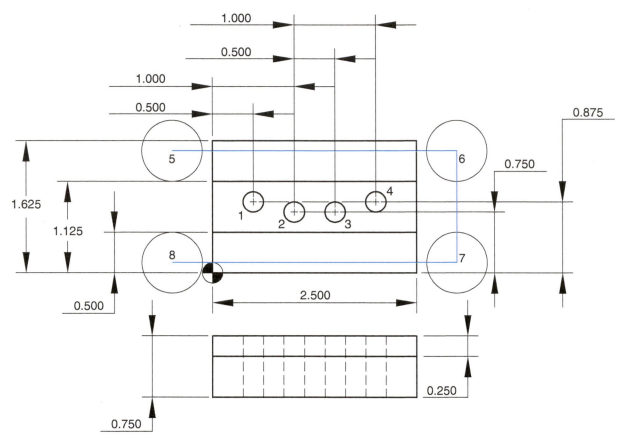

Clamp Block

Coordinate Table

Position	X	Y	Z
1			
2			
3			
4			
5			
6			
7			
8			

Tool List

Order	Operation	Tool	Tool station	rpm	Feed (ipm)
1	centerdrill 4 holes	#3 centerdrill	1	3000	4.5
2	drill (4) 0.250" holes	0.25" drill	2	1600	5.0
3	mill steps	0.75" end mill	3	2000	4.0

Program

O1001 (Clamp Block)
(Centerdrill)

1. N005 G17 G_____ G80 G40 G49
2. N010 G90 G54 S_____ M3 T1 M6
3. N015 G_____ X _____ Y _____ (Position 1)
   N020 M1
4. N025 G_____ H1 Z.2
   N020 M8
5. N025 G_____ Z–.2 F4.5
6. N030 G_____ Z.2
7. N035 X _____ Y _____ (Position 2)
8. N040 _____ _____ _____
9. N045 G00 _____
10. N050 _____ _____ (Position 3)
11. N055 _____ _____ _____
12. N060 _____ _____
13. N065 _____ _____ (Position 4)
14. N070 _____ _____ _____
15. N075 _____ _____ M9
16. N080 G91 _____ Z0 G49
    (1/4 Drill)
17. N085 G17 G20 G80 G_____ G49
18. N090 G90 G54 S_____ M3 T2 M6
19. N095 G_____ X _____ Y _____ (Position 1)
    N100 M1
20. N105 G43 _____ _____
    N110 M8
21. N115 G_____ Z–.850 F_____
22. N120 G_____ Z.2
23. N125 X_____ Y_____ (Position 2)
24. N130 _____ _____ _____
25. N135 G00 _____
26. N140 _____ _____ (Position 3)
27. N145 _____ _____ _____
28. N150 _____ _____
29. N155 _____ _____ (Position 4)
30. N160 _____ _____ _____
31. N165 _____ _____ M9
32. N170 G91 _____ Z0 _____
    (.75 End mill)
33. N175 G17 G20 G80 G40 _____
34. N180 G_____ G54 S_____ M3 T_____ M6
35. N185 G_____ X_____ Y_____ (Position 5)
    N190 M1
36. N195 G43 H____ Z.2
    N200 M8

37. N205 G_____ Z _____ F_____
38. N210 X_____ Y_____ (Position 6)
39. N215 G_____ X_____ Y_____ (Position 7)
40. N220 _____ _____ _____ (Position 8)
     N225 M9
41. N230 G91 G_____ Z0 G_____
42. N235 G91 _____ _____
     N240 M30

2. Complete the following programming activity.
    A. Determine the coordinate positions shown on the slot plate print. Place these coordinate values in the coordinate table. Application of trigonometry will be required to determine various cutter positions. Note: Cutter Position 1 has a value of X3.125 Y3.625.
    B. Use the information from above to determine the codes, spindle speed, and feed rate missing in the program provided.

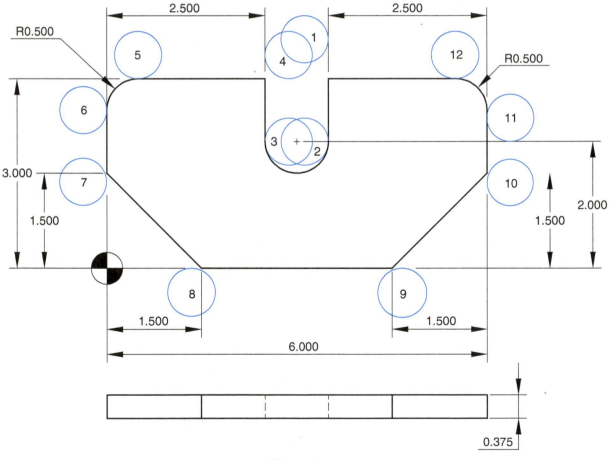

Slot Plate

Coordinate Table

Position	X	Y	Z
1			
2			
3			
4			
5			
6			
7			
8			
9			
10			
11			
12			
13			

Tool List

Order	Operation	Tool	Tool station	rpm	Feed (ipm)
1	mill contour	0.75 end mill	1	350	4.0

Program

O1002 (Slot Plate)
(.75 End Mill)

1. N005 G17 G20 G80 G40 _____
2. N010 G90 G54 S_____ M3 T1 M6
    N015 G00 X3.125 Y3.625 (Position 1)
    N020 M1
    N025 G00 H1 Z.2
3. N030 M_____
    N035 G01 Z–.375 F4.0
4. N040 Y_____ (Position 2)
5. N045 G_____ X2.875 R_____ (Position 3)
6. N050 G_____ Y_____ (Position 4)
7. N055 X_____ (Position 5)
8. N060 G_____ X_____ Y_____ I0 J_____ (Position 6)
9. N065 G_____ Y_____ (Position 7)
10. N070 X_____ Y_____ (Position 8)
11. N075 X_____ (Position 9)
12. N080 X_____ Y_____ (Position 10)
13. N085 _____ (Position 11)
14. N090 G_____ X_____ Y_____ I_____ (Position 12)
15. N095 G_____ X_____ (Position 13)
    N100 M9
16. N105 G_____ G28 Z0 G49
17. N110 G_____ G28 X0 Y0
18. N115 M_____

3. Complete the following programming activity.
   A. Determine the coordinate positions shown on the trip plate print. Place these coordinate values in the coordinate table. Application of trigonometry will be required to determine various tool positions. Note: A dashed triangle is given as help to compute distances needed in determining cutter positions.
   B. Calculate speeds and feeds based on aluminum material, and place in the operation sheet.
   C. Use the information from above to determine the codes, spindle speed, and feed rate missing in the program provided.

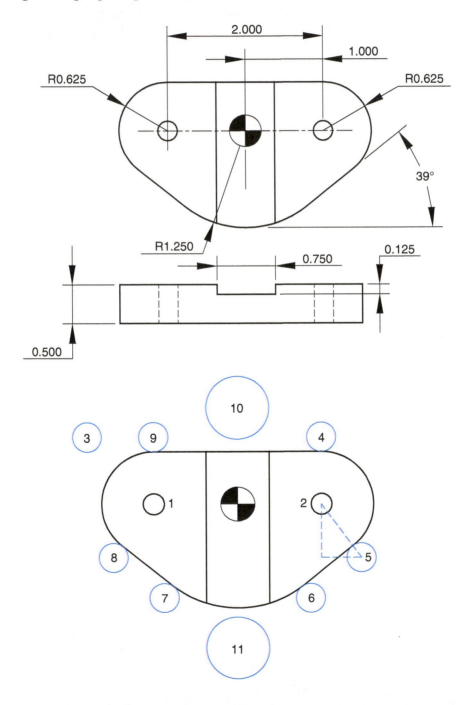

Coordinate Table

Position	X	Y	Z
1			
2			
3			
4			
5			
6			
7			
8			
9			
10			
11			

Tool List

Order	Operation	Tool	Tool station	rpm	Feed (ipm)
1	centerdrill (2) holes	#3 centerdrill	1		
2	drill (2) holes	0.25" drill	2		
3	mill contour	0.25" end mill	3		
4	mill slot	0.75" end mill	4		

Program

O1002 (Trip Plate)
(#3 Centerdrill)

1. N005 G17 G_____ G80 G40 G49
2. N010 G90 G54 S_____ M3 T1 M6
   N015 G00 X–1 Y0 (Position 1)
3. N020 _____
4. N025 G00 H1 Z_____
   N030 M8
5. N035 G01 Z–.2 F_____
   N040 G00 Z.2
6. N045 X_____ (Position 2)
7. N050 G01 Z–.2 F_____
   N055 G00 Z.2 M9
8. N060 G91 _____ Z0 G49
   (1/4" Drill)
9. N065 G17 G20 G80 G40 G_____
10. N070 G90 G54 S_____ M3 T2 M_____
    N075 G00 X–1. Y0 (Position 1)
    N080 M1
    N085 G00 H2 Z.2
    N090 M8
11. N095 G01 Z–.575 F_____
    N100 G00 Z.2
    N105 X1. (Position 2)
12. N110 G01 Z–.575 F_____
    N115 G00 Z.2 M9
    N120 G91 G28 Z0 G49
    (.250 End Mill)
    N125 G17 G20 G80 G40 G49

13. N130 G90 G54 S_____ M3 T3 M6
14. N135 G00 X–1.5 Y_____ (Position 3)
    N140 M1
    N145 G00 H3 Z.2
15. N150 _____
16. N155 G01 Z–.5 F_____
    N160 X1 (Position 4)
17. N165 G_____ X_____ Y_____ J–.875 (Position 5)
18. N170 G_____ X_____ Y_____ (Position 6)
19. N175 G_____ X_____ I–1.073 J_____ (Position 7)
20. N180 G_____ X_____ Y_____ (Position 8)
21. N185 G_____ X–1. Y_____ I.585 J_____ (Position 9)
    N190 G00 Z.2 M9
    N195 G91 G28 Z0 G49
    (.75 End Mill)
    N200 G17 G20 G80 G40 G49
22. N205 G90 G54 S_____ M3 T4 M6
23. N210 G00 X_____ Y1.125 (Position 10)
    N215 M1
24. N220 G_____ H4 Z.2
    N225 M8
25. N230 G_____ Z_____ F_____
    N235 Y–1.75 (Position 11)
    N240 M9
    N245 G91 G28 Z0 G49
    N250 G91 G28 X0 Y0
    N255 M30

Chapter 10
Canned Cycles for Machining Centers

Objectives

Information in this chapter will enable you to:

- Recognize the various components of a canned cycle.
- Define the various components of a canned cycle.

Technical Terms

boring canned cycle	peck drill canned cycle	spotface, counterbore
canned cycles	reaming canned cycle	canned cycle
counterboring canned cycle	simple drill cycle	tapping canned cycle

Shortening Programs

Canned cycles are a method of shortening the length of programs while performing such operations as drilling, centerdrilling, reaming, spotfacing, counterboring, tapping, and boring. The canned cycle code gives information to the controller about how the cycle is to be conducted, while the program lines following these instructions give the tool positions for these instructions. Program examples of the various canned cycles are given at the end of this chapter. All canned cycles have the following in common:

- Positioning the tool to an X, Y location with a rapid traverse move.
- Positioning the tool to a clearance or R-plane with a rapid traverse move.
- Feeding the tool in the Z axis to a specified depth.
- Removing the tool from the hole to a clearance or R-plane.

The various words that can be used with canned cycles include the following:

- **G80–G89.** Occurs every cycle. These are canned cycle codes.

- **N.** Occurs every cycle. This is the line number.
- **X.** Occurs every cycle. This is the X position of the hole.
- **Y.** Occurs every cycle. This is the Y position of the hole.
- **Z.** Occurs every cycle. This is the Z position of the hole.
- **R.** Occurs every cycle. This is the depth or bottom of the hole.
- **F.** Occurs every cycle. This is the feed rate.
- **G98.** Occurs every cycle. This is the rapid out position from the hole to the initial plane.
- **G99.** Occurs every cycle. This is the rapid out position from the hole to the R-plane.

> ### Note
>
> The G98 and G99 codes tell the control where to return along the Z axis after the hole is machined. The G98 returns the tool along the Z axis to the initial plane (last Z position prior to the canned cycle command). This Z position is a height that is safely above any obstructions such as clamps or workpiece planes. The G99 code is the R-plane (rapid plane) that is defined in the canned cycle command. This height is usually 0.200″ or 0.100″ above part zero, but can be any distance the programmer desires. These codes should be placed before or after the canned cycle command on the same line.

- **P.** Occurs with G82 and G89. This is the pause length at the bottom of the hole, measured in thousandths of a second (milliseconds). For example, P750 equals 0.75 seconds.
- **Q.** Occurs with G83. This is the amount of peck distance.

The chart in **Figure 10-1** shows various attributes of the common machining center canned cycles.

Common Machining Center Canned Cycles				
G-Code	**Function**	**Z axis motion**	**Depth**	**Retract Motion**
G81	Drill, counterdrill	Feed		Rapid out
G82	Spotface, counterbore	Feed	Dwell	Rapid out
G83	Peck drill	Peck feed		Rapid out
G84	Tap	Feed	Reverse spindle	Feed out
G85	Bore, ream	Feed		Feed
G86	Bore–drag line	Feed	Spindle stop	Rapid out
G89	Counterbore	Feed	Pause, spindle stop	Rapid out

Figure 10-1. This chart lists the attributes of common canned cycles.

Canned Cycles

The G81 cycle is the *simple drill cycle*. It is a simple cycle that feeds the tool to a specified depth then rapids out of the hole. Codes used in this cycle, as shown in **Figure 10-2**, include:

- **G99.** Tool retracts to R-plane after each hole is drilled
- **G81.** States a drill canned cycle
- **X__Y__.** Tool rapids to drilling position
- **Z.** Depth of the drilling cycle
- **R.** R-plane (position from Z part zero to which tool retracts)
- **F.** Drilling feed rate in inch/min or mm/min

The G82 cycle is the *spotface, counterbore canned cycle*. It is similar to G81, except the tool pauses at the bottom of the hole while the spindle rotates. This is done so that tool pressure is relieved. It is important to maintain close tolerance on hole depth. Codes used in this cycle, as shown in **Figure 10-3**, include:

- **G99.** Tool retracts to R-plane after each hole is machined
- **G82.** States a counterbore canned cycle
- **X___Y___.** Tool rapids to machining position
- **Z.** Counterbore depth
- **R.** R plane (position from Z part zero to which tool retracts)
- **F.** Feed rate of tool in in/minute or mm/minute
- **P.** Dwell time (pause at bottom of hole in milliseconds)

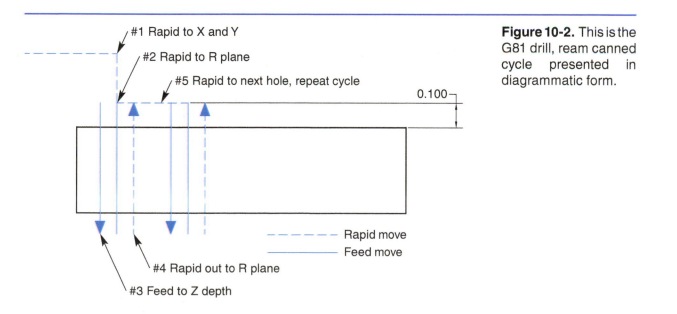

#1 Rapid to X and Y

#2 Rapid to R plane

#5 Rapid to next hole, repeat cycle

0.100

\- - - - - - Rapid move

———— Feed move

#4 Rapid out to R plane

#3 Feed to Z depth

Figure 10-2. This is the G81 drill, ream canned cycle presented in diagrammatic form.

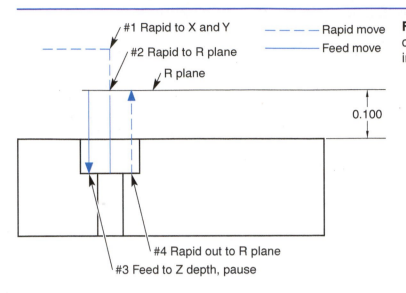

#1 Rapid to X and Y — — — Rapid move
#2 Rapid to R plane ——— Feed move
R plane

0.100

#4 Rapid out to R plane
#3 Feed to Z depth, pause

Figure 10-3. This is the G82 spotface, counterbore canned cycle presented in diagrammatic form.

The G83 cycle is the *peck drill canned cycle*. This cycle is used to remove chips when drilling deep holes. This is done so that the drill flutes do not become clogged and break the drill. A peck depth is calculated, and the cycle works in this manner: rapid to R-plane, feed down to Q value, rapid to R-plane, rapid down to last Z value less 0.010, feed Q value, repeat until Z depth is reached, and rapid up. Codes used in this cycle, as shown in **Figure 10-4**, include:

- **G99.** Tool retracts to R-plane after each hole is drilled
- **G83.** States peck drill canned cycle
- **X___Y___.** Tool rapids to drilling position
- **Z.** Hole depth
- **Q.** Peck depth, distance drill feeds down each time until desired hole depth is reached (rapids to R-plane after each peck)
- **R.** R-plane (position from Z part zero to which tool retracts)
- **F.** Feed rate of tool in in/min or mm/min

The G84 cycle is the *tapping canned cycle*. This cycle consists of a rapid move to a location, a rapid move to the R-plane, feeding down to depth, reversing the spindle direction, and feeding back to the R-plane or initial plane. Codes used in this cycle, as shown in **Figure 10-5**, include:

- **G99.** Tool retracts to R-plane after each hole is tapped
- **G84.** States tapping canned cycle
- **X___Y___.** Tool rapids to tapping position
- **Z.** Depth of tapped hole
- **R.** R-plane (position from Z part zero to which tool retracts)
- **F.** Feed rate of tap in in/min or mm/min

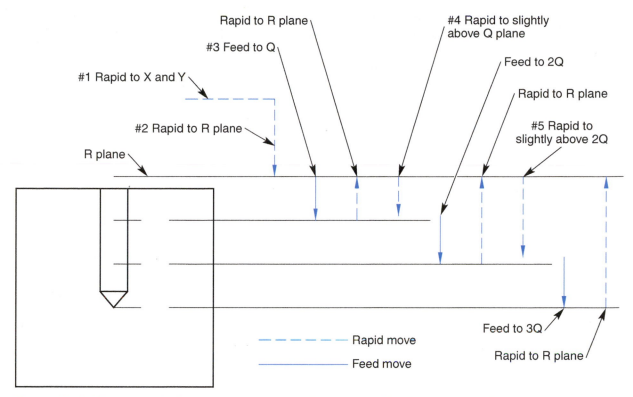

Figure 10-4. This is the G83 peck drill canned cycle presented in diagrammatic form.

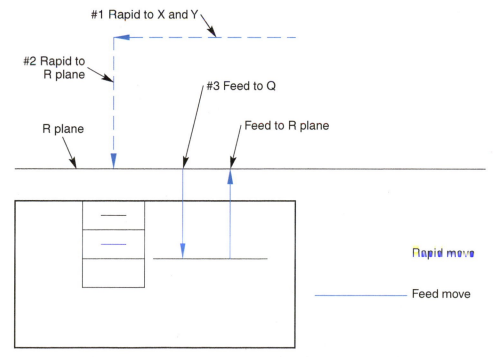

Figure 10-5. This is the G84 tapping canned cycle presented in diagrammatic form.

> **Note**
>
> The tap retracts at the feed rate.

The feed rate is important because it involves the lead of the tap. The formula for determining feed rate is as follows:

Tapping feed rate $=$ rpm $\times$ lead of tap

Lead of tap is 1/thread per inch
Example: 1/4-20 UNC tap, aluminum material

$$\text{rpm} = \frac{4(50)}{0.25}$$

$$\text{rpm} = \frac{200}{0.25}$$

$$\text{rpm} = 800$$

Tapping feed rate $=$ 800×0.05

Tapping feed rate $=$ 40

The G85 *reaming canned cycle* is very similar to the G81 cycle. However, with the G85 cycle, the tool *feeds* out of the workpiece instead of *rapiding* out. Codes used in this cycle, as shown in **Figure 10-6**, include:

- **G99.** Tool retracts to R-plane after each hole is reamed
- **G85.** States reaming canned cycle

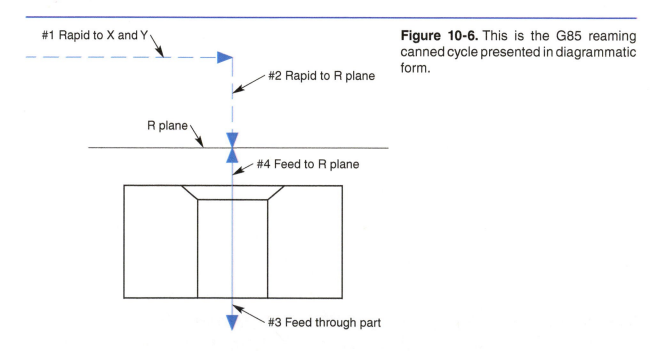

Figure 10-6. This is the G85 reaming canned cycle presented in diagrammatic form.

- **X___Y___.** Tool rapids to reaming position
- **Z.** Depth of reamed hole
- **R.** R-plane (position from Z part zero to which tool retracts)
- **F.** Feed rate of tool in in/minute or mm/minute.

> ### Note
> The tool retracts from the hole at the feed rate.

The G86 *boring canned cycle* is used when a drag line (mark) is allowed in the hole caused by the tool rapiding out of the hole when the spindle stops. Codes used in this cycle, as shown in **Figure 10-7,** include:

- **G99.** Tool retracts to R-plane after each hole is bored
- **G86.** States boring canned cycle
- **X___Y___.** Tool rapids to boring position
- **Z.** Depth of bored hole
- **R.** R-plane (position from Z part zero to which tool retracts)
- **F.** Feed rate of tool in in/minute or mm/minute

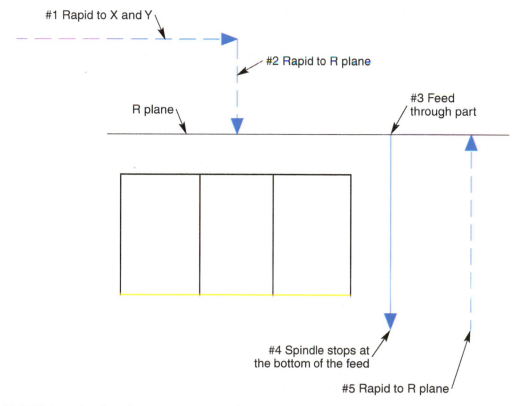

Figure 10-7. This is the G86 boring canned cycle presented in diagrammatic form.

> ## Note
> The spindle rotation stops at the hole depth before retracting.

The G89 *counterboring canned cycle* is almost the same as the G82 cycle. The G89 cycle is used with a boring bar; and the spindle stops before rapiding out of the hole. This cycle is used to produce a smooth hole to a precise depth. Codes used in this cycle, as shown in **Figure 10-8**, include:

- **G99.** Tool retracts to R-plane after each hole is counterbored
- **G89.** States counterboring canned cycle
- **X___Y___.** Tool rapids to hole position
- **Z.** Counterbore depth
- **R.** R-plane (position from Z part zero to which tool retracts)
- **F.** Feed rate of tool in in/minute or mm/minute
- **P.** Dwell (pause) at counterbore depth (P in milliseconds)

> ## Note
> The tool feeds out to the R-plane.

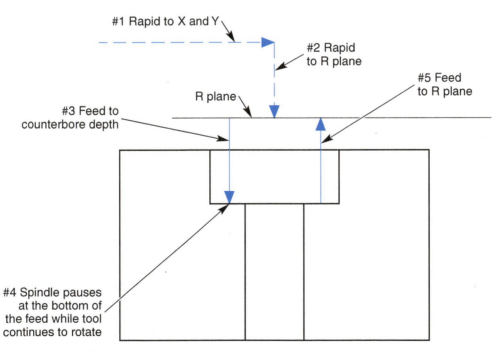

Figure 10-8. This is the G89 counterboring canned cycle presented in diagrammatic form.

Canned Cycle Programs

This section contains two examples of programs using canned cycles presented in this chapter.

Simple Drilling Program—G81 and G82

This example will cover G81 and G82 canned cycles. Refer to **Figure 10-9.**

	Tool List			
Tool No.	Operation	Tool description	Speed (rpm)	Feed (ipm)
1	Drill thru (4) No. 8 holes	No. 8 HSS drill	2200	6
2	Counterbore (4) .312 × .250DP	.312–2-flute HSS end mill	1100	5

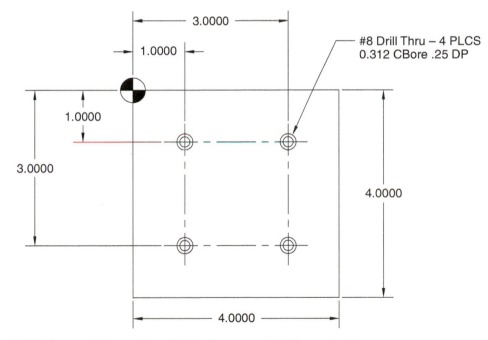

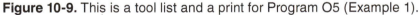

Figure 10-9. This is a tool list and a print for Program O5 (Example 1).

O5 (Block)

N10 (X0Y0 is located in the upper-left corner of the workpiece)

N20 (Z0 is located on the top of the workpiece)

N30 (Tool 1 is a No. 8 drill)

N40 (Tool 2 is a 0.312 end mill)

N50 (Workpiece is mild steel 4×4×1)

(Drill 4 holes thru)

N60 G17G20G80G40G49	*Default block*
N70T1M6 (No. 8 drill)	*Tool change*
N80 G90G54S2200M3	*Absolute, fixture offset, spindle speed, CW rotation*
N90 G00X1.0Y-1.0	*Rapid to hole 1*
N100 M1	*Optional stop*
N110 G43H1Z.2	*Tool length compensation, offset 1, R-plane 0.2"*
N120 M8	*Coolant on*
N130 G99G81X1.0Y-1.0Z-1.1R.2F6	*Drill hole 1*
N140 Y-3	*Drill hole 2*
N150 X3	*Drill hole 3*
N160 Y-1.0	*Drill hole 4*
N170 G80	*Cancel canned cycle*
N180 G00Z.2M9	*Rapid to R-plane, coolant off*
N190 G91G28Z0G49	*Rapid to Z home, cancel tool length compensation*

(Counterbore 4 holes 0.312 × 0.25 deep)

N200 G17G20G80G40G49	*Default block*
N210T2M6 (0.312 end mill)	*Tool change*
N220 G90G54S1100M3	*Absolute, fixture offset, spindle speed, CW rotation*
N230 G00X1.0Y-1.0	*Rapid to hole 1*
N240 M1	*Optional stop*
N250 G43H2Z.2	*Tool length compensation, offset 2, R-plane 0.2"*
N260 M8	*Coolant on*
N270 G99G82X1.0Y-1.0Z-.25R.2F5.P1500	*Counterbore hole 1*
N280 Y-3.	*Counterbore hole 2*
N290 X3.0	*Counterbore hole 3*
N300 Y-1.0	*Counterbore hole 4*
N310 G80	*Cancel canned cycle*
N320 G00Z.2M9	*Rapid to R-plane, coolant off*
N330 G91G28Z0G49	*Rapid to Z home, cancel tool length compensation*
N340 G91G28X0Y0	*Return to X and Y home position*
N350 M30	*End and rewind program*

Peck Drilling Program—G83 and G85

This program example will cover canned cycles G83 and G85. See **Figure 10-10**. This program requires three setups to machine the part. The first setup requires an aluminum workpiece that is 2 3/4″× 5″× 1 1/4″. The profile is milled with this setup. See **Figure 10-11**. The second setup requires the workpiece to be flipped and milled to a thickness of 0.875″. See **Figure 10-12**. Finally, the third setup, which is shown in the print, is where all the holes are machined. Refer back to Figure 10-10.

First setup

1. Workpiece must be raised a minimum of 15/16″ above the vise jaws.

2. Part zero is set in the middle of the workpiece 1/8″ from the left end of the workpiece. See sketches A and B.

3. Part Z0 is the top of part.

Second setup

1. Workpiece is flipped over to machine excess thickness. See sketch C.

2. Part zero is the middle of the workpiece at the left edge.

3. Part Z0 is set 0.875″ above parallels.

Third setup

1. Workpiece zero is located in the upper-left hand corner of the workpiece.

2. Part Z0 is top of part.

O6
N10 (Part X0Y0 changes for each setup, see setup directions)
N20 (Z0 is located on the top of the workpiece)
N30 (Tool 1 is a 0.5″ HSS end mill)
N40 (Tool 2 is a 3″ carbide face mill)
N50 (Tool 3 is a #3 centerdrill)
N60 (Tool 4 is a #7 drill)
N70 (Tool 5 is a 1/4-20 UNC tap)
N80 (Tool 6 is a 23/32″ HSS drill)
N90 (Tool 7 is a 0.75″ reamer)
N100 (Tool 8 is a 1″ drill)
N110 (Tool 9 is a 1.25″ boring bar)
(Milling outside profile—first setup)

N120 G17G20G80G40G49	*Default block*
N130T1M6 (0.5″ end mill)	*Tool change*
N140 G90G54S1500M3	*Absolute, fixture offset, spindle speed. CW rotation*

Tool List

Tool No.	Operation	Tool description	Speed (rpm)	Feed (ipm)
1	Mill profile	.5 HSS end mill	1500	6
2	Face mill thickness	3″ carbide face mill	600	10
3	Centerdrill 6 holes	#3 centerdrill	3000	2
4	Peck drill (4) #7 holes	#7 (.201) drill	1500	6
5	Tap (4) 1/4-20 holes	1/4-20 UNC tap	400	20
6	Pre-drill ream hole	23/32″ drill	900	5
7	Ream hole	.75″ ream	450	3.5
8		1″ drill	700	9
9		1.25″ boring bar	500	2

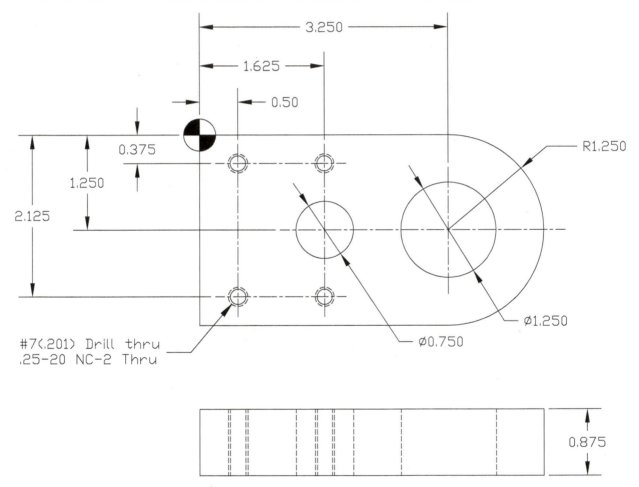

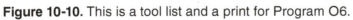

Figure 10-10. This is a tool list and a print for Program O6.

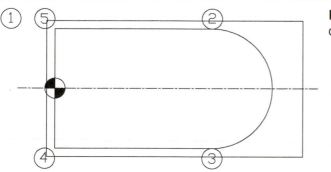

Tool path example 2 — Setup 1

Figure 10-11. The tool path for the outside contour is completed in Setup 1.

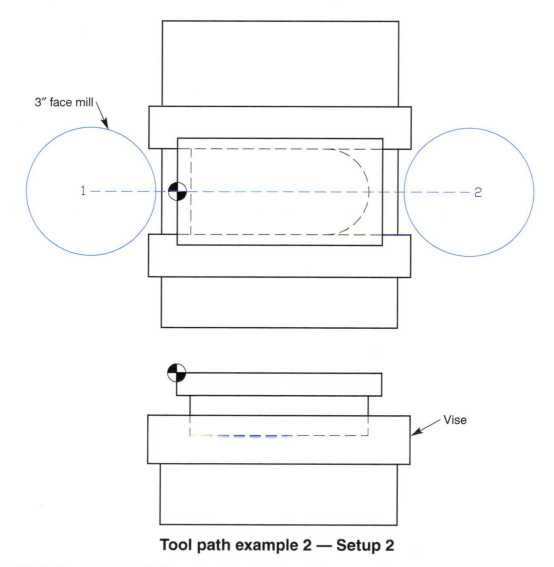

Tool path example 2 — Setup 2

Figure 10-12. The workpiece is flipped and ready to be milled to the final thickness.

N150G00X-.75Y1.5	*Rapid to position 1*
N160M1	*Optional stop*
N170G43H1Z-.875	*Rapid to depth*
N180M8	*Coolant on*
N190G1X3.25F6	*Linear feed to position 2*
N200G02Y-1.5R1.5	*Circular arc to position 3*
N210G1X-.25	*Linear feed to position 4*
N220Y1.5	*Linear feed to position 5*
N230G00Z.2M9	*Rapid to R-plane, coolant off*
N240G91G28Z0G49	*Rapid to Z home, cancel tool length compensation*
N250M0 (Perform second setup)	*Programmed stop*
(Face mill workpiece to thickness—second setup)	
N260G17G20G80G40G49	*Default block*
N270T2M6 (3" face mill)	*Tool change*
N280G90G54S600M3	*Absolute, fixture offset, spindle speed, CW rotation*
N290G00X-2Y0	*Rapid to position 1*
N300M1	*Optional stop*
N310G43H2Z0	*Rapid to Z0 position*
N320M8	*Coolant on*
N330G1X7	*Feed to position 2*
N340G00Z.2M9	*Rapid to R-plane, coolant off*
N350G91G28Z0G49	*Rapid to Z home, cancel tool length compensation*
N360M0 (Perform third setup)	
(Drill, tap, and bore holes—third setup)	
N370G17G20G80G40G49	*Default block*
N380T3M6 (Centerdrill)	*Tool change*
N390G90G54S3000M3	*Absolute, fixture offset, spindle speed, CW rotation*
N400G00X.5Y-.375	*Rapid to hole 1*
N410M1	*Optional stop*
N420G43H3Z.1	*Rapid to R-plane*
N430M8	*Coolant on*
N440G99G81X.5Y-.375Z-.375R.1F2	*Drill hole 1*
N450Y-2.125	*Drill hole 2*
N460X1.625	*Drill hole 3*
N470Y-1.25	*Drill hole 4*
N480Y-.375	*Drill hole 5*
N490X3.25Y-1.25	*Drill hole 6*
N500G80	*Cancel canned cycle*
N510G00Z1.M9	*Rapid above part, coolant off*
N520G91G28Z0G49	*Rapid to Z home, cancel tool length compensation*
N530G17G20G80G40G49	*Default block*

N540T4M6 (#7 drill)	*Tool change*
N550G90G54S3000M3	*Absolute, fixture offset, spindle speed, CW rotation*
N560G00X.5Y-.375	*Rapid to hole 1*
N570M1	*Optional stop*
N580G43H4Z.1	*Rapid to R-plane*
N590M8	*Coolant on*
N600G99G83X.5Y-.375Z-.95Q.3R.1F6	*Drill hole 1*
N610Y-2.125	*Drill hole 2*
N620X1.625	*Drill hole 3*
N630Y-.37 5	*Drill hole 5*
N640G80	*Cancel canned cycle*
N650G00Z2.M9	*Rapid 2" above part, coolant off*
N660G91G28Z0G49	*Rapid to Z home, cancel tool length compensation*
N670G17G20G80G40G49	*Default block*
N680T5M6 (1/4-20UNC Tap)	*Tool change*
N690G90G54S400M3	*Absolute, fixture offset, spindle speed, CW rotation*
N700G00X.5Y-.375	*Rapid to hole 1*
N710M1	*Optional stop*
N720G43H5Z.250	*Rapid 0.25"above part*
N730M8	*Coolant on*
N740G99G84X.5Y-.375Z-.95R.1F20	*Tap hole 1*
N750Y-2.125	*Tap hole 2*
N760X1.625	*Tap hole 3*
N770Y-.375	*Tap hole 5*
N780G80	*Cancel canned cycle*
N790G00Z1.M9	*Rapid above part, coolant off*
N800G91G28Z0G49	*Rapid to Z home, cancel tool length compensation*
N810G17G20G80G40G49	*Default block*
N820T6M6 (23/32" HSS drill)	*Tool change*
N830G90G54S900M3	*Absolute, fixture offset, spindle speed, CW rotation*
N840G00X1.625Y-1.25	*Rapid to hole 4*
N850M1	*Optional stop*
N860G43H6Z.1	*Rapid to R-plane*
N870G1Z-1.125F5	*Drill hole 4*
N880G00Z.1M9	*Rapid to R-plane, coolant off*
N890G91G28Z0G49	*Rapid to reference point return, cancel tool length*
N900G17G20G80G40G49	*Default block*
N910T7M6 (0.75" reamer)	*Tool change*
N920G90G54S450M3	*Absolute, fixture offset, spindle speed, CW rotation*
N930G00X1.625Y-1.25	*Rapid to hole 4*
N940M1	*Optional stop*

N950G43H6Z.1	*Rapid to R-plane*
N960G1Z-1.F3.5	*Ream hole 4*
N970G00Z.1M9	*Rapid to R-plane, coolant off*
N980G91G28Z0G49	*Rapid to reference point return, cancel tool length*
N990G17G20G80G40G49	*Default block*
N1000T8M6 (1" HSS drill)	*Tool change*
N1010G90G54S700M3	*Absolute, fixture offset, spindle speed, CW rotation*
N1020G00X3.25Y-1.25	*Rapid to hole 6*
N1030M1	*Optional stop*
N1040G43H6Z.1	*Rapid to R-plane*
N1050M8	*Coolant on*
N1060G1Z-1.2F9	*Drill hole 4*
N1070G00Z.1M9	*Rapid to R-plane, coolant off*
N1080G91G28Z0G49	*Rapid to Z home, cancel tool length compensation*
N1090G17G20G80G40G49	*Default block*
N1100T9M6 (1.25" boring bar)	*Tool change*
N1110G90G54S500M3	*Absolute, fixture offset, spindle speed, CW rotation*
N1120G00X3.25Y-1.25	*Rapid to hole 6*
N1130M1	*Optional stop*
N1140G43H6Z.1	*Rapid to R-plane*
N1150M8	*Coolant on*
N1160G99G86X3.25Y-1.25Z-.9R.1F2	*Bore hole 6*
N1170G80	*Cancel canned cycle*
N1180G00Z2.M9	*Rapid 2" above part, coolant off*
N1190G91G28Z0G49	*Rapid to Z home, cancel tool length compensation*
N1200G91G28X0Y0	*Return to X and Y home position*
N1210M30	*End and rewind program*

Summary

Canned machining cycles reduce the length of programs. This is done by providing information about a machining operation and repeating that information at additional locations.

Chapter Review

Answer the following questions. Write your answers on a separate sheet of paper.

1. What code is used to cause a tool to rapid out of a hole to an initial plane?
2. What does the letter *R* mean?
3. What canned cycle is used for tapping?
4. What canned cycle is used for spotfacing?
5. Explain what peck drilling does.
6. What does the letter *Q* mean?
7. What type of machining operations can be accomplished with a G81 command?
8. Why is the letter *P* used?
9. What is the usual height of an R-plane?
10. What retract motion is used in a G86 canned cycle?

Activities

1. Read the following program and determine which canned cycle should be used to shorten the program. Rewrite the program with the canned cycle.

O0001
N10 (X0Y0 is located in the upper-left corner of the workpiece)
N20 (Z0 is located on the top of the workpiece)
N30 (Tool 1 is a No. 8 drill)
N40 T1M6 (No. 8 drill)
N50 G90G54S2200M3
N60 G00X1.0Y–1.0
N65 M8
N70 G43H01Z.2
N80 G01Z–.25F3.0
N90 G00Z.2
N100 G01Z–.5F3.0
N110 G00Z.2
N120 G01Z–.75F3.0
N130 G00Z.2
N140 G01Z–1.0F3.0
N150 G00Z.2M9
N160 G91G28Z0G49

2. Obtain programs from several companies and identify the canned cycles used, explain the purpose of each cycle, and describe the components of each command.

A well-designed fixture holds the workpiece securely and accurately, while allowing quick part changes. (Tibor Machine Products)

N10G20G99G40
N20G96S800M3
N30G50S4000
N40T0100M8
N50G00X3.35Z1.25T0101
N60G01X3.25F.002
N70G04X0.5
N80X3.35F.05
N90G00X5.0Z0T0101
01111
N10G20G99G40
N20G96S800M3
N30G50S4000
N40T0100M8
N50G00X3.35Z1.25T0101
N60G01X3.25F.002
N70G04X0.5
N80X3.35F.05

Chapter 11
Turning Centers

Objectives

Information in this chapter will enable you to:

* Name two types of turning centers.
* List various components found on turning centers.
* List various operations performed on turning centers.
* List methods of workholding on turning centers.

Technical Terms

bar pullers	grooving	tailstock
bed	headstock	tap
boring	live tooling	tapping
chip conveyors	machine control unit	threading
collet chuck	(MCU)	three-jaw chuck
contour turning	part catchers	tool presetters
drilling	parting	turning
facing	reamer	turning centers
gang-style tool holder	reaming	turret

Types of Turning Centers

Turning centers can be grouped into five machine categories:

* Two-axis turning center
* Four-axis turning center
* Twin spindle turning center
* Vertical turning center
* Gang turning center

Although these machines may appear to be different, the directions of tool motion are the same for all machines. For all of these turning centers, the X-plus motion is away from the centerline of the machine and the X-minus motion is toward the centerline of the machine. The Z-minus motion is toward the chuck end of the machine, while Z-plus motion is away from the workpiece. See **Figure 11-1**.

Two-Axis Turning Center

The two-axis turning center is the most popular turning center. It is usually found as a slant bed style, **Figure 11-2**. The universal slant bed style has a tailstock attached to the machine while the chucking style is a smaller machine without a tailstock. The slant style bed permits chips to easily slide down into the chip bed.

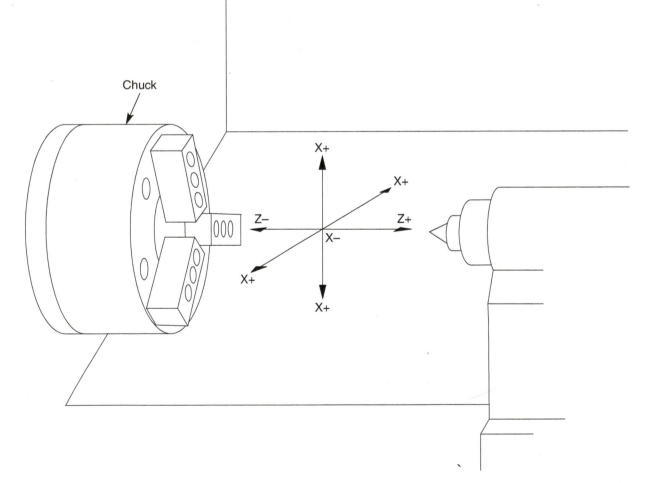

Figure 11-1. Coordinate directions for turning centers are X– is toward the axis and X+ is away from the axis. Z– is toward the chuck and Z+ is away from the chuck.

Figure 11-2. This slant bed two-axis turning center has a live tailstock. (Hardinge)

Four-Axis Turning Center

The four-axis turning center has an additional tool turret. See **Figure 11-3**. This allows machining to occur using two tools at the same time. The turrets are located on each side of the workpiece. Each turret is operated by its own program. A good programmer can reduce machining time significantly by using two turrets to machine a workpiece.

Twin Spindle Turning Center

Twin spindle turning centers, **Figure 11-4**, are available in two types of configuration. These configurations consist of spindles opposing each other at both ends of the machine or two spindles side by side.

Figure 11-3. Four-axis turning centers permit operations to be performed on both sides of the workpiece, significantly reducing machining time. (Mazak)

Vertical Turning Center

Vertical turning centers are used for large, heavy workpieces with the spindle mounted in a vertical position, **Figure 11-5**. Tool direction remains the same as on horizontal machines. A bar feed mechanism cannot be adapted to this type of machine.

Gang Turning Center

Gang turning centers are small machines that have their tools mounted on a table rather than a turret, **Figure 11-6**. While tool setup time can be longer than other types of machines, there is no tool changing time once the setup is complete. Gang turning centers are commonly used for bar work.

CNC Components

CNC turning centers contain many of the same components as the conventional lathe. Turning centers include the headstock, tailstock, bed, spindle, tool holder (turret), and carriage (slide).

Figure 11-4. Foot pedals are used to operate the chucks on this twin spindle turning center with opposing spindles.

Figure 11-5. This vertical turning center includes milling capability. (O-M Ltd.)

Figure 11-6. Multiple tool setup on a gang turning center. The workpiece is held in a collet chuck.

The *headstock* contains the spindle and spindle drive motor. See **Figure 11-7**. The spindle drive motor is a variable speed motor. Various holding devices, such as chucks, are mounted to the headstock. The *tailstock* supports long and heavy work. The tailstock can be programmed to function with commands within a program or can be manually operated. It is not found on all machines and is usually used when the work length exceeds three times the work diameter.

The *three-jaw chuck* may be hydraulically or pneumatically operated. See **Figure 11-8**. The hydraulic chuck has a high degree of clamping power. It is equipped with either hard or machinable soft jaws. Foot pedals control the clamping and unclamping of these jaws.

Headstock

Tailstock

Figure 11-7. This headstock is on a benchtop turning center. The chuck must be tightened with a key. This tailstock unit is on a slant bed turning center.

Figure 11-8. The jaws of this three-jaw hydraulic chuck are machined to fit snugly around the workpiece. (Goss & DeLeeuw Machine Company)

The *collet chuck*, **Figure 11-9**, may be hydraulically or pneumatically operated. It is usually used with bar feeders and is programmable.

The *turret* is the most common tool holder, **Figure 11-10**. It can have an 8-, 10-, or 12-tool capacity. Tools are mounted either upright or upside down. Most turrets rotate in either direction taking the shortest path to select the desired tool. They can, however, be programmed to travel in a specific direction to avoid a collision when tool changes are made close to the workpiece and tools are of different lengths.

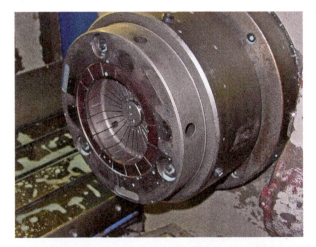

Figure 11-9. A collet chuck, such as this one found on a twin spindle turning center, will clamp the workpiece at nearly every point around its circumference.

Figure 11-10. The bidirectional design of this 12-station turret permits the turret to decide the direction of rotation for maximum efficiency. The direction can also be programmed by the operator.

A *gang-style tool holder* has tools mounted on a table, **Figure 11-11**. It requires a more difficult setup, but is most efficient when tool changing, since a turret is not involved and less time is needed to change to another tool. This type of tool holder is used on small CNC machines.

The *bed* supports the major components of a machine. It can be flat or slanted. See **Figure 11-12**.

The *machine control unit (MCU)* contains a computer that stores and runs programs. See **Figure 11-13**. It has a control panel containing switches, buttons, and a monitor screen.

Bar pullers, **Figure 11-14**, are mounted in the turret. They are used to clamp and pull bars of stock through the spindle for machining.

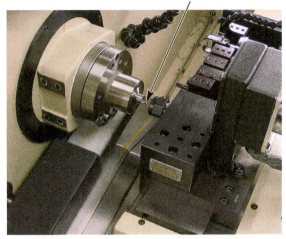

Figure 11-11. This gang tooling is being used on a Hardinge lathe with the end mill forming internal slots in a steel workpiece.

Figure 11-12. The slant bed on this turning center causes chips to fall to the chip bin.

Figure 11-13. A Fanuc controller on a turning center. Membrane covers eliminate problems caused by dirt and moisture.

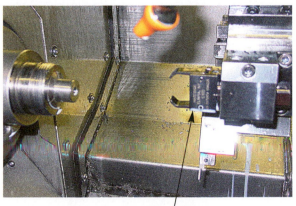

Figure 11-14. To grip and pull bar stock from this collet chuck, a bar puller (pullout fingers) can be mounted on the turning center turret.

Part catchers are programmable devices used to catch parts as they are cut off the bar stock, **Figure 11-15**.

Live tooling is the term for tools that can rotate while mounted in the turret. This allows the workpiece to be milled, drilled, and tapped in various positions. **Figure 11-16** shows examples.

Tool presetters automatically measure tools held in a turret, **Figure 11-17**. The measurements are used for locating purposes.

Chip conveyors are used to remove chips from a machine. See **Figure 11-18**. The chips are then dumped into a hopper or other container.

Figure 11-15. This is a typical part catcher on a two-axis turning center.

Part catcher

Drill head

Slot   Milling head

Figure 11-16. The drill head is drilling the last of three holes on the face of the aluminum part. The milling head is using a four-flute endmill to cut external slots in the flange of this part. (Sandvik Coromant)

Tool presetter

Figure 11-17. Setting tool offset length on turning tool with a tool presetter. (Haas)

Chip conveyor

Figure 11-18. Chip conveyors can be started manually or programmed to start at certain intervals.

Basic Turning Center Operations

Facing removes material from the end of the workpiece. This results in a surface that is square with the workpiece centerline, **Figure 11-19**. It is generally the first operation performed on a workpiece. Cutting is accomplished along the X axis.

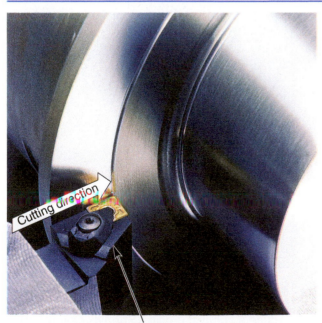

Cutting direction

Toolholder

Figure 11-19. Facing operation being performed on a cast iron workpiece using an indexable toolholder. (Sandvik Coromant)

Turning reduces the diameter of the workpiece. This is the most frequently performed operation. See **Figure 11-20**. The tool travel takes place in the Z axis. This operation can be performed using standard G-codes or special canned cycles. Straight turning and taper turning are accomplished using rough and finish passes.

Contour turning involves various operations such as chamfering, turning, recessing, grooving, tapering, and radius forming. These operations are combined to generate a specific shape to the workpiece. See **Figure 11-21**.

Boring is an internal turning operation, **Figure 11-22**. It is used to finish a hole to size and remove any imperfections. Boring can be used to finish holes when a good finish is required and a reamer of the correct size is not available. Boring enlarges a previously drilled or cast hole. Holes can be straight, tapered, or contoured.

Grooving can be done as an external or internal process. See **Figure 11-23**. Grooving is accomplished using a tool that is fed perpendicular to the workpiece centerline to a specified depth. Grooving is often used to provide relief for a threading operation.

Parting is used to remove the finished workpiece from the bar stock, **Figure 11-24**. The tool is fed perpendicular to the bar stock until the workpiece is separated from the stock.

Drilling, **Figure 11-25**, is used to produce a hole in the workpiece. Common drills include centerdrills, twist drills, spade drills (for large holes), and carbide insert drills. The tools are fed in the Z axis, usually on center (X0).

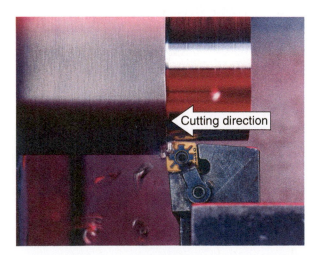

Figure 11-20. Indexable toolholder performing a turning operation. (Carboloy)

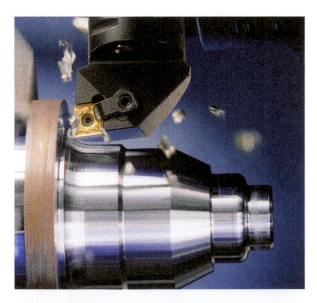

Figure 11-21. Contour turning involves cutting arcs, curves, tapers, chamfers, and shoulders to produce a specific profile. (Sandvik Coromant)

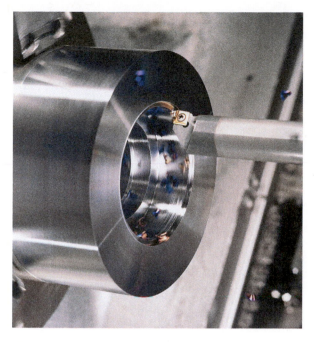

Figure 11-22. The largest size boring bar available should be used to avoid tool deflection when performing a boring operation such as this. (Kennametal)

Figure 11-23. Grooves can be internal or external, as shown. This plunge cutting technique makes the grooves perpendicular to the centerline of the workpiece. (Carboloy)

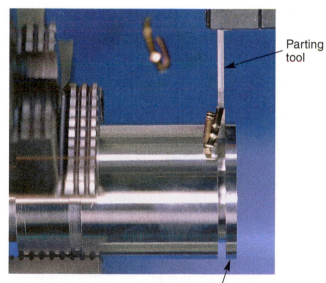

Parting tool

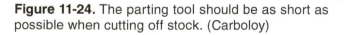

Finished part being cut off

Figure 11-24. The parting tool should be as short as possible when cutting off stock. (Carboloy)

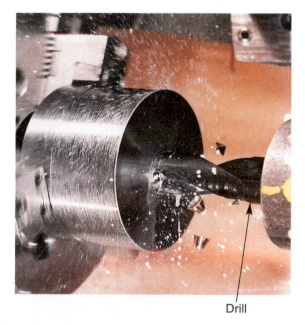

Drill

Figure 11-25. The high-speed drill advances into the part as the part is rotated. The coolant is being applied to reduce heat and wash away chips. (Kennametal)

Reaming finishes a hole to size, **Figure 11-26**. This is done with a multiflute tool called a *reamer*. Drilling, and sometimes, boring are performed prior to reaming.

Threading forms internal or external helical grooves. See **Figure 11-27**. This operation uses a specially ground tool that conforms to a specific type of thread desired.

Tapping is used to produce internal threads. See **Figure 11-28**. It is performed with a multiflute tool called a *tap*.

Figure 11-26. Reaming a plastic workpiece using a straight flute reamer held in an extension holder.

Plastic workpiece   Reamer   Extension holder

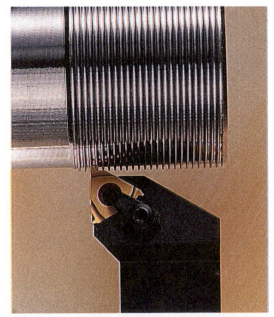

Figure 11-27. Forming an external thread with an indexable threading tool. (Carboloy)

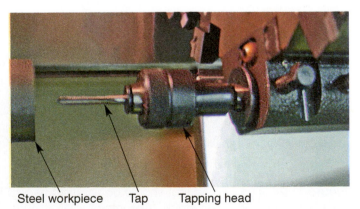

Steel workpiece   Tap   Tapping head

Figure 11-28. This tap is held in tapping head in preparation for tapping a steel workpiece.

Summary

There are five types of turning centers. They differ in appearance, but are common in tool motion. There are various components on CNC turning centers that are also common to conventional lathes.

Special equipment used with turning centers includes part catchers, bar feeders, chip conveyors, measuring devices, and live tooling. The most basic operations performed on a turning center include facing, turning, drilling, reaming, grooving, parting, threading, boring, and tapping.

Chapter Review

Answer the following questions. Write your answers on a separate sheet of paper.

1. Name two types of turning centers.
2. List five components found on turning centers.
3. Name five operations performed on a turning center.
4. List three methods of holding or supporting work on a turning center.
5. Write the name of each basic turning operation shown.

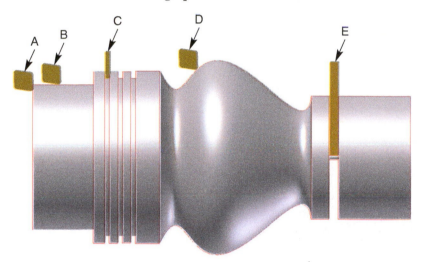

Activities

1. Search the internet for suppliers of CNC turning centers. Print the images of five different turning centers, including one vertical turning center. The images should be large enough to allow only one image per page. Draw arrowheads showing the X, Y, and Z axes on the images of the vertical turning center and two of the horizontal turning centers. Label the arrows and place a plus and minus at the correct ends of each arrow. On the two remaining images, circle and label the following components, if visible: headstock, 3-jaw chuck, tailstock, turret station, cutting tool, and control panel.

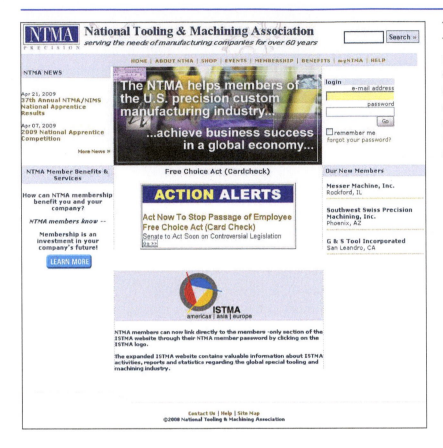

The National Machining and Tooling Association helps members of the U.S. precision custom manufacturing industry achieve business success in a global economy through advocacy, advice, networking, information, programs, and services. www.ntma.org

Chapter 12
Turning Center Tools, Inserts, Speeds, and Feeds

Objectives

Information in this chapter will enable you to:

- Interpret the components of a toolholder identification system.
- Select a toolholder for a specific operation.
- Describe the ANSI insert identification system.
- List various machining terms.
- Calculate revolutions per minute, or rpm.
- Use charts to determine machining data.

Technical Terms

coarse feeds	finishing cuts	positive rake inserts
cutting speed	machine screw	qualified
depth of cut	clamping	roughing cuts
external toolholder	multiple clamping	rpm (revolutions per
identification system	negative rake inserts	minute)
feed	pin lock clamping	top clamping
fine feeds		

Insert Toolholders

Insert toolholders are manufactured from heat-treated steel. The tool end is machined with a pocket that supports a shim and the insert. See **Figure 12-1**. In most instances, a carbide shim is fastened to the bottom of the pocket. This shim, along with the cutting insert, helps absorb the cutting forces. The shim also guarantees a flat surface for the insert to rest. The pocket seat is machined to accept a neutral, positive, or negative insert. This makes the holder a neutral, positive, or negative rake holder.

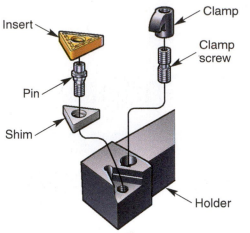

Insert

Pin

Shim

Clamp

Clamp
screw

Holder

Figure 12-1. These are the various parts of an insert toolholder. (Sandvik Coromant)

Toolholders are labeled *qualified* when the dimensions from the tip of the insert to the side of the toolholder and the end of the toolholder are within a tolerance of ±0.003″.

There are four general methods used to fasten the insert to the holder. *Top clamping* uses a clamp and clamp screw to fasten an insert. *Pin lock clamping* uses an eccentric pin to hold the insert in place. *Multiple clamping* employs both a clamp and eccentric pin, while *machine screw clamping* uses a fine thread button head screw.

Tool System Selection

The type of tool selected for a machining operation should be based on whether the operation to be performed is external or internal. External operations affecting the choice of toolholders include facing, turning, profiling, grooving (necking), cutoff, threading, and turning slender parts. See **Figure 12-2**. Internal operations include boring, facing, profiling, threading, and grooving (necking). The system selected also includes the method of holding the insert itself. After selecting the tool system, the selection of the actual toolholder is done. When making this decision, the following factors are considered:

- **Dimension (size) of the tool.** The tool must fit the machine turret.

- **Tool's lead angle.** This is the angle of cutting edge when it enters the workpiece.

- **Insert point angle.** This is the included angle of the insert tip.

- **Versatility.** The toolholder must have the ability to cut the part contour, which includes curves and angles.

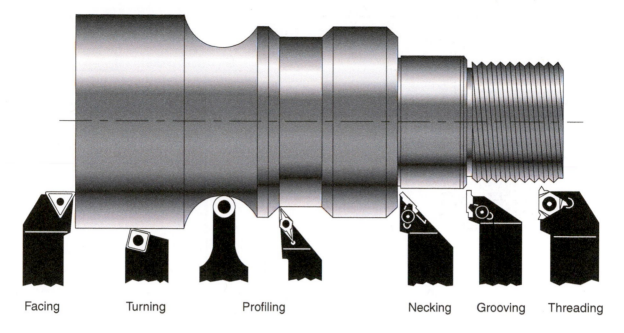

| Facing | Turning | Profiling | Necking | Grooving | Threading |

Figure 12-2. External operations on a turning center include those shown here. (Sandvik Coromant)

Tooling manufacturers' catalogs are quite extensive and contain important technical data about tooling. Detailed pictures and descriptions of the many types and styles of carbide insert holders are furnished, as well as details about the inserts used with these holders.

Figure 12-3 shows various styles of external toolholders and their applications using some inserts.

Toolholder Identification System

To identify toolholders, manufacturers use an identification system that conforms to the standards set by the American National Standards Institute (ANSI). The system covers the method of holding the insert, insert shape, toolholder style, toolholder rake, hand of tool (right or left), toolholder shank size, insert size, and toolholder qualification.

Figure 12-4 is a chart of the ANSI standards used to identify external toolholders. In addition to the *external toolholder identification system* shown, there are identification systems for internal tools (boring, threading, grooving), cartridges (removable heads that attach to holders), and special applications, such as multidirectional turning (MDT).

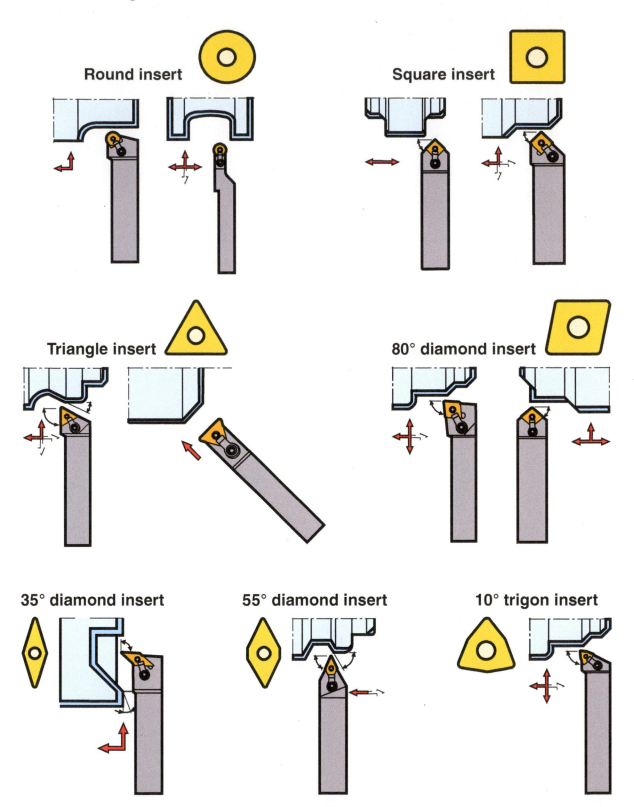

Round insert

Square insert

Triangle insert

80° diamond insert

35° diamond insert

55° diamond insert

10° trigon insert

Figure 12-3. These are some examples of the many external toolholders used with various inserts. (Carboloy)

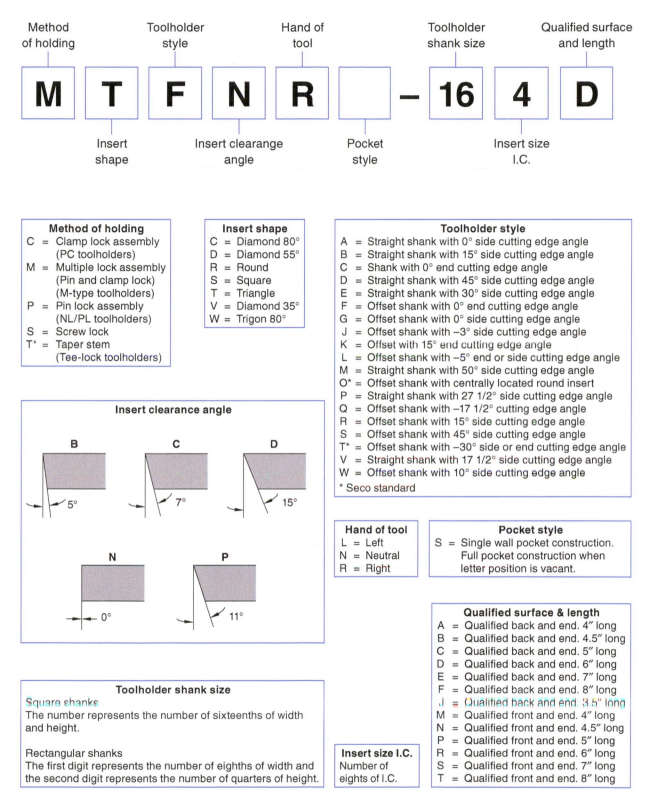

Figure 12-4. External toolholders identification system developed by the American National Standards Institute (ANSI). (Carboloy)

Cutting Inserts

Almost all turning centers use indexable (disposable) cutting inserts. These inserts are made of various materials and are manufactured in a number of shapes, **Figure 12-5**. A feature of these types of inserts is to provide an insert with several cutting edges. When one edge becomes broken or worn, the insert can be indexed (rotated) in the toolholder until all the edges are used. The insert is then discarded and a new one mounted on the toolholder. Advantages of using indexable inserts are the following:

- The ability to rapidly change the inserts without removing the tooling from the machine

- The cost of regrinding the tool is eliminated

- The various geometries used on inserts provide many choices for chip control and finishes

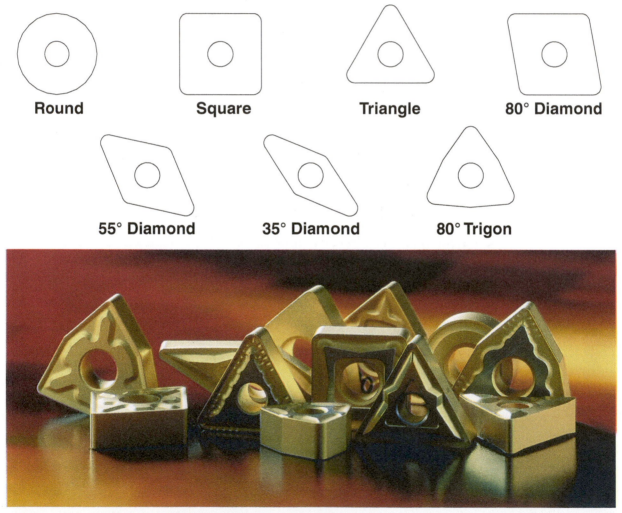

Figure 12-5. Indexable cutting inserts. A—Common insert shapes. B—Examples of inserts. (Carboloy)

Negative and Positive Rake

Inserts and insert tool holders are made with a neutral, negative, or positive rake for cutting different materials. See **Figure 12-6**. *Negative rake inserts* are used for machining the majority of steels and cast irons. These inserts work especially well on rough and interrupted cuts. Negative rake inserts have a high use on 200 and 300 series stainless steels and high temperature alloys.

Positive rake inserts help cut down on chatter or part bending. Positive inserts are used when machining long, slender, or thin-walled workpieces. Positive rake inserts are used on soft steels and nonferrous metals. The cutting force required is much smaller than when using negative inserts.

Insert Identification System

The insert identification system is a standardized method of describing an insert. Each identifying name consists of a series of up to nine letters and numbers that identify the shape of the insert, the dimensions of the insert, and any special conditions or considerations. For example:

DNCG-432E

- **D.** *shape* (55° diamond)
- **N.** *relief angle* (0°)
- **C.** *insert tolerance class* (IC±0.0005", thickness, ±0.001")
- **G.** *insert type* (with hole and chipbreaker on both faces)
- **–.** optional hyphen (not a position) used to separate insert shape information from dimension information.
- **4.** *insert size* (4/8" or 1/2")
- **3.** *insert thickness* (3/32")
- **2.** *cutting point radius* (1/32")
- **E.** *cutting edge condition* (rounded cutting edges)

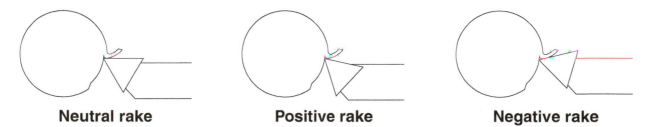

Neutral rake   **Positive rake**   **Negative rake**

Figure 12-6. These triangular inserts are positioned to provide a neutral, positive, and negative rake, depending on the type of material the tool is cutting.

Shape

As noted, the first position in the insert identification system uses a letter to describe the insert shape. Standard shapes and their identifying letters are listed:

- **R.** Round
- **V.** 35° Diamond
- **E.** 75° Diamond
- **D.** 55° Diamond
- **T.** Triangle
- **S.** Square
- **A.** 85° Parallelogram
- **M.** 86° Diamond
- **W.** 80° Trigon
- **C.** 80° Diamond
- **K.** 55° Parallelogram
- **P.** Pentagon
- **O.** Octagon
- **L.** Rectangle

Relief angle

The second position in the insert identification system shows the *relief angle*, also known as *clearance*. This angle may be a combination of the insert and toolholder. The relief angle can vary from 0°, which is used with negative rake holders, to 25°.

Tolerances

The third position in the insert identification system indicates two values that control the indexability of the insert. These are the IC value and the insert thickness value.

Type

The fourth position in the insert identification system refers to the kind of insert design and how it is held in toolholder. Inserts may have various combinations of mounting holes and chipbreakers, as shown in **Figure 12-7**.

Size

The fifth position is a one- or two-digit number calculated from one of two formulas, depending on the shape of the insert. If the insert is a regular polygon or a diamond, the number of 32nds (inserts smaller than 1/4") or 8ths (inserts with 1/4" IC or greater) of the diameter of the IC. The shape of the insert is specified in the first position of the insert identification.

Insert Types
A = With hole, no chipbreaker
E = Smaller than 1/4″ IC without hole
G = With hole and chipbreaker on both faces
M = With hole and chipbreaker on one face only
R = Without hole, chipbreaker on one face only
T = C-lock hole with chipbreaker on one face only
W = C-lock hole, no chipbreaker

Figure 12-7. An example of insert type descriptions and identification letters for use in the fourth position of the insert identification system.

If the insert is a rectangle or parallelogram, the first digit is the width in 8ths of an inch and the second digit is the length in 4ths of an inch.

Regular polygon or diamond	
IC is less than 1/4″	IC is greater than or equal to 1/4″
Number of 32nds in the diameter of the IC (1/8″ IC = 4/32, the size of the insert is 4)	Number of 8ths in the diameter of the IC (3/4″ IC = 6/8, the size of the insert is 6)

Rectangle or parallelogram	
First digit	Second digit
Number of 8ths in the width of the insert (3/8″ wide, the first size digit is 3)	Number of 4ths in the length of the insert (1/2″ long = 2/4, the second size digit is 2)

Thickness

The sixth position in the insert identification system is a one- or two-digit number representing the thickness of the insert. The number comes from the quantity of 32nds (inserts smaller than 1/4″) or 16ths (inserts with 1/4″ IC or greater).

Point radius

The seventh position in the insert identification system is a number describing the radius of the insert's point (or points). See **Figure 12-8**.

Cutting edge condition

The eighth position in the insert identification system is a single letter showing the condition of the insert's cutting edges, **Figure 12-9**.

Manufacturer's option

The ninth position in the insert identification system may be used as an option by the insert manufacturer.

Point Radius, Flats
0 = Sharp to 1/256″ Radius
0.5 = 1/128″ Radius
1 = 1/64″ Radius
2 = 1/32″ Radius
3 = 3/64″ Radius
4 = 1/16″ Radius
6 = 3/32″ Radius
8 = 1/8″ Radius

Figure 12-8. Identification numbers for various point radius values.

Cutting Edge Condition
F = Sharp cutting edges
E = Rounded cutting edges
T = Chamfered cutting edges
S = Chamfered and rounded cutting edges
K = Double chamfered cutting edges
P = Double chamfered and rounded cutting edges
Cutting edge condition is dependent on size and shape of the insert.

Figure 12-9. Identification numbers for various cutting edge conditions.

Cutting Speed, RPM, and Feed for Turning Centers

Cutting speed is a measurement of the distance (in feet or meters) that the circumference of the work passes the cutting tool in one minute. Cutting speed is given in surface feet per minute (sfpm) or surface meters per minute (smpm). It is based on the type of material being machined and the cutting tool material being used. There are many charts that give cutting speed values of many materials based on the cutting tool material being used. The range of cutting speed will vary due to other factors. These factors include depth of cut, tool design, work size and shape, setup rigidity, machine rigidity, cutting fluids, machine power, and the finish required.

The number of spindle rotations in one minute is expressed as *rpm (revolutions per minute)*. It is calculated using the following formula:

$$\text{rpm} = 3.82 \times \text{sfpm} \div D$$

$$\text{rpm} = \text{revolutions per minute}$$

$$\text{sfpm} = \text{surface feet per minute}$$

$$D = \text{diameter of the workpiece (inches)}$$

Feed is the distance the tool moves along the work during each revolution of the spindle. It is stated in thousandths of an inch. Some tables provide ranges of feeds to use for various depths of cuts. Feed values will vary based on depth of cut, as well as the type of material being cut and the surface finish required.

The two feed categories are coarse and fine. *Coarse feeds* are used for rough cutting and generally softer materials. *Fine feeds* are used for finishing and for cutting harder materials. *Depth of cut* is the distance the tool is fed perpendicular into the workpiece. It determines the amount of material being removed.

Roughing cuts are used to remove excess material without regard to surface finish. *Finishing cuts* are used to size the workpiece and provide a good surface finish. Several roughing cuts are sometimes needed to remove a large amount of material. **Figure 12-10** shows the recommended speeds, feeds, and depth of cuts for various materials.

Recommended Speeds, Feeds, and Depths of Cut Using Carbide Inserts			
Material	**Depth of Cut** inches	**Feed per Rev.** inches	**Cutting Speed** feet/minute
Aluminum	0.005–0.015	0.002–0.005	700–1000
	0.020–0.090	0.005–0.015	450–700
	0.100–0.200	0.015–0.030	300–450
	0.300–above	0.030–0.090	100–200
Brass (Bronze)	0.005–0.015	0.002–0.005	700–800
	0.020–0.090	0.005–0.015	600–700
	0.100–0.200	0.015–0.030	500–600
	0.300–above	0.030–0.090	200–400
Cast Iron	0.005–0.015	0.002–0.005	350–450
	0.020–0.090	0.005–0.015	250–350
	0.100–0.200	0.015–0.030	200–250
	0.300–above	0.030–0.090	75–150
Machine Steel	0.005–0.015	0.002–0.005	700–1000
	0.020–0.090	0.005–0.015	550–700
	0.100–0.200	0.015–0.030	400–550
	0.300–above	0.030–0.090	150–300
Tool Steel	0.005–0.015	0.002–0.005	500–750
	0.020–0.090	0.005–0.015	400–500
	0.100–0.200	0.015–0.030	300–400
	0.300–above	0.030–0.090	100–300
Stainless Steel	0.005–0.015	0.002–0.005	375–500
	0.020–0.090	0.005–0.015	300–375
	0.100–0.200	0.015–0.030	250–300
	0.300–above	0.030–0.090	75–175

Figure 12-10. Recommended feeds, speeds, and depth of cuts for various materials when using carbide inserts.

There is a relationship between tool nose radius and feed rate that results in various surface finishes. **Figure 12-11** shows this relationship.

Summary

There are many different styles of toolholders for both external and internal tools. There are also several methods of fastening the insert to the toolholder. The toolholder identification system was developed for classifying the many types of toolholders.

Cutting inserts are designated by shape, relief angle, tolerances, type, size, thickness, point radius, and cutting edge condition, using a system developed by ANSI.

Cutting speed is the distance (feet or meters) that the circumference of the work passes the cutting tool in one minute. Spindle rotation is measured in rpm, or revolutions per minute. Feed is the distance the tool moves along the work during each revolution of the spindle. Depth of cut is the distance the tool is fed perpendicular into the workpiece.

The finish obtained on a workpiece is influenced by the feed rate and nose radius of the cutting tool.

Finish versus Nose Radius

Nose Radius inches	Feed Rate inches/rev	Finish rms
0.015	0.005	125
	0.0025	63
	0.0012	32
0.031	0.01	125
	0.005	63
	0.0026	32
0.046	0.015	125
	0.076	63
	0.003	32

Figure 12-11. Effect of nose radius and feed rate on surface finish.

Chapter Review

Answer the following questions. Write your answers on a separate sheet of paper.

1. Name three methods of fastening inserts to holders.
2. What criteria are used to select a toolholder?
3. What does the first position in the toolholder identification system specify?
4. What does the third position in the toolholder identification system specify?
5. What does the seventh position in the toolholder identification system specify?
6. What does the eighth position in the toolholder identification system specify?
7. What does the term *qualify* mean?
8. What style toolholder is used to form a 45° chamfer?
9. What style toolholders use triangular inserts?
10. List six shapes of carbide inserts.
11. Which two types of relief angles are used on inserts?
12. What does the third position in the ANSI insert identification system relate to?
13. What does *IC* mean?
14. What does an IC value of 4 mean?
15. What does an insert thickness value of 3 mean?
16. What does a point radius value of 0 mean?
17. List four factors that can influence cutting speed.
18. Define the term *feed*.
19. To obtain a 63 finish with a 0.031 tool nose radius, what feed rate must be used?

Activities

1. Using the formula for rpm and the table of recommended speeds, feeds, and depths of cut (Figure 12-10), calculate the rpm for the values given in the following table. Copy the table on a separate sheet of paper and place the answers in the empty cells.

Material	Depth of Cut	Cutting Speed	Part Diameter	rpm
Cast Iron	0.1	low value	1.75	
Aluminum	0.015	high value	3	
Machine Steel	0.05	low value	2.5	
Tool Steel	0.7	low value	1.5	
Stainless	0.02	high value	2.25	

This view shows the large number of tools that can be mounted on the indexer of a turning center.

Chapter 13
Programming Process for Turning Centers

Objectives

Information in this chapter will enable you to:

- Record items to be included on a tool list.
- Determine cutting conditions.
- Read and understand a turning center program.
- Write turning center programs using the format presented.

Technical Terms

diameter programming	manufacturing instruction sheet (MI)	program manuscript radius programming
graphical prove-out		setup sheet

Programming Sequence

Although this section is similar to Chapter 7, *Programming Process for Machining Centers*, it serves as a reminder of the importance of adequate preparation when writing a CNC program. It also contains information specific to turning centers.

Analyze the Print

If a print is not dimensioned in absolute values, it becomes necessary to label and mathematically identify datum points (centerline of the workpiece, and usually, the right end of the workpiece). The location of all features and machining positions should be labeled using numbers or letters to identify positions. **Figure 13-1** shows a part print and its conversion to a working print.

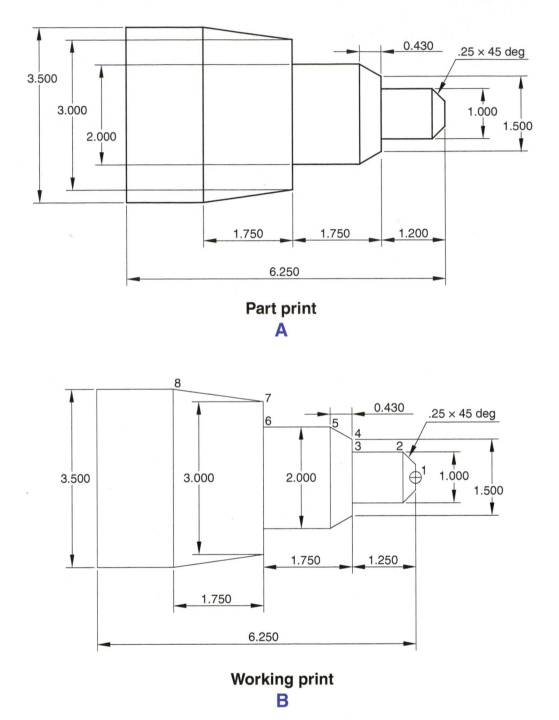

Part print

A

Working print

B

Figure 13-1. Converting a part print to a working print. A—Part print of a multidiameter workpiece. B—Working print showing various tool positions on the workpiece.

Determine the Sequence of Operations

Determine the logical sequence of machining operations to take place. This procedure is based on the industrial experience of the programmer. The information is usually placed on a procedure sheet called a *manufacturing instruction sheet (MI)* or manufacturing methods sheet, **Figure 13-2**. Methods sheets may vary significantly from one company to another. However, the elements found on a methods sheet usually include the part name, part number, machine name, material specifications, fixture or holding device identification, and columns containing sequence numbers and descriptions of each sequence.

Select a Turning Center

There are many factors to consider when selecting a turning center. These factors include the following:

- Availability
- Size of the workpiece
- Travel limits
- Horsepower
- Turret stations
- Operational cost
- Number of workpieces needed
- Holding device

Manufacturing Instruction Sheet			
Part Number	**Part Name**	**Material**	**Machine**
Sequence	**Description**		

Figure 13-2. An example of a typical instruction sheet used by a programmer. Companies usually design their own sheets.

Prepare a Tool List

The sequence of cutting tools to be used in the program must be listed. The tool list can greatly help in organizing a program. Not all companies use a tool list. Some companies include tool information within the part program instead. Items that may be found on tool lists include the following:

- Part number
- Part name
- Program number
- Operation number
- Tool sequence number (the tool number [T] used in the program)
- Tool description (type of tool, such as drill or boring bar)
- Tool diameter
- Tool offset number (last two digits with T-word, specifying offset register that contains tool length and diameter offset value)
- Tool nose radius
- Tool length
- Tool tip (identifying one of the nine different tool tip configurations necessary when applying tool nose radius compensation)
- Carbide inserts (may be included with tool description)
- Surface feet per minute (sfpm)
- Spindle speed (rpm)
- Feed (ipr)

Prepare Setup Sheet and Program Manuscript

A *setup sheet* lists the holding device, such as chuck, special chuck jaws, chuck pressure, and tailstock. A sketch is sometimes included to clarify the machine setup procedure.

The methods sheet is used to develop a program manuscript. The *program manuscript* is the written CNC program containing the preparatory and miscellaneous codes. It also contains the axis information, feed rates, and spindle speeds.

Load and Verify Program

Various input data or procedures may be used to load a program, such as MDI, tape, disc, or downloading. To verify the program, *graphical prove-out* can be used or a dry run (with or without the workpiece) can be conducted.

Perform single block run without workpiece

This verification can be performed in addition to a graphical prove-out, especially with new programs, to make sure no crash will occur during rapid travel in various directions or during tool changing. It is simply another method to avoid mistakes in programming. Graphical prove-out may be sufficient to determine if a job can be run without damaging the tool, machine, or operator.

Perform single block run with workpiece

This is the last verification step and is performed using the override switches to control the speed of the program. Care should be taken when approaching the workpiece so that no crashes occur; tool retract paths also should be checked to make sure crashes are avoided. Offsets can be adjusted to control size dimensions.

Run the Part

Inspect the part with a part print and make necessary part size and finish adjustments. Rerun the workpiece if necessary and recheck the part for accuracy.

Diameter and Radius Programming

Dimensions in programming can be specified in two ways. If the diameter is specified, it is called *diameter programming*; if the radius is specified, it is called *radius programming*. See **Figure 13-3**. Radius programming is found on older controls. Diameter and decimal programming is the choice on new controls.

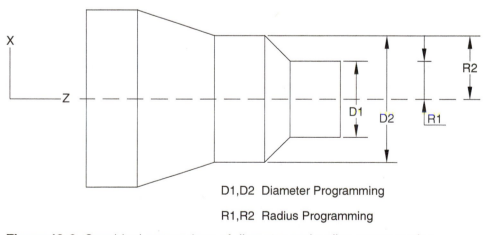

D1,D2 Diameter Programming

R1,R2 Radius Programming

Figure 13-3. Graphical comparison of diameter and radius programming

Program Format

The following examples show the basic program structure (programming format) that is used for turning center programs. Remember, formats differ with different controllers or with different companies and programmers. The format that will be used here is Fanuc-based.

Program Start

N10G20G99G40	*Inch mode, feed per revolution, tool nose radius compensation cancel*
N20G96S300M3	*Constant surface speed control, surface speed value, spindle direction control*
N30G50S3000	*Set maximum spindle speed, spindle speed*
N40T0100M8	*Tool change, coolant on*
N50G00X3.Z3.T0101	*Rapid to tool change position, tool offset*

Tool Return

N90G00X3.Z3.T0900M9	*Rapid to tool change position, cancel tool offset, coolant off*
N100M5	*Spindle off*
N110M1	*Optional stop*

Tool Change

N130G96S400M3	*Constant surface speed control, spindle speed, Spindle direction (CW)*
N140G50S3000	*Set maximum spindle speed, spindle speed*
N150T0200M8	*Tool change, coolant on*

Program End

N250G00X3.Z3.T0200M9	*Rapid to tool change position, cancel tool offset, coolant off*
N260G28U0W0	*Return to home position*
N270M30	*Program end*

Note

Some companies do not use line numbers, except where tools are introduced. This is done to save on program memory. Also, sometimes tool changes are made at home with rapid moves directly to the start of cut rather than at a tool change position.

Turning—Example 1

Refer to the **Figure 13-4** as you study this sample program.

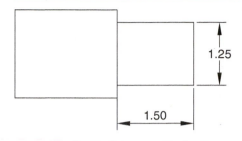

Note: Part No. O1111
1 3/8" Dia. × 3 7/8" Lg. Stock
Part Zero is right end of workpiece

1.25

1.50

Figure 13-4. Part print for sample program O1111.

O1111	*Program number*
N10G20G99G40	*Inch mode, feed per revolution, tool node radius compensation cancel*
N20G96S800M3	*Constant surface speed control, surface speed value, spindle direction (CW)*
N30G50S4000	*Set maximum spindle speed, spindle speed*
N40T0100M8	*Tool change, coolant on*
N50G00X2.Z2.T0101	*Rapid to tool change position, tool offset*
N60X1.6Z.010	*Rapid to clearance plane*
N70G97S600	*Cancel constant surface speed, spindle rpm*
N80G01X-.064F.010	*Linear feed, rough face stock at .010 ipr*
N90Z.050	*Retract from workpiece at feed rate*
N100G00X1.250	*Rapid to first rough cut position*
N110G01Z-1.490	*Linear feed first rough turn*
N120X1.3Z-1.47	*Retract from cut at feed rate*
N130G00Z.100	*Rapid to Z axis clearance plane*
N140G00X1.165	*Rapid to second rough cut position*
N150G01Z-1.49	*Second rough turn*
N160X1.215Z-1.47	*Retract from cut at feed rate*
N170G00X2.Z2.T0100M9	*Rapid to tool change position, cancel tool offset, coolant off*
N180M5	*Spindle off*
N190M1	*Optional stop*
N200G96S400M3	*Constant surface speed control, spindle speed, spindle*
N210G50S4000	*Set maximum spindle speed, spindle speed*
N220T0200M8	*Tool change, coolant on*
N230G00X1.365Z0T0202	*Rapid to clearance plane in X axis, tool No. 2 offset 2*
N240G01X-.062F.007	*Finish face end of stock, .007 ipr feed*
N250Z.1	*Retract to Z clearance plane at feed rate*
N260G00X1.125	*Rapid to start of finish turn*
N270G01Z-1.5	*Finish turn 1.125 diameter*
N280X1.6	*Finish face of the 1.125 diameter shoulder*
N290G00X2.Z2.T0200M9	*Rapid to tool change position, cancel tool offset, coolant off*
N300G28U0W0	*Return to home position*
N310M30	*Program end*

Turning—Example 2

Refer to the **Figure 13-5** as you study this sample program. Determine the cutting tools needed for the job, and determine the machining cut conditions. Calculate the necessary tool positions needed for part programming.

O2222	*Program*
G20G99G40	*Inch mode, feed per revolution, tool nose radius compensation cancel*
G96S500M4	*Constant surface speed control, surface speed value, spindle direction (CCW)*
G50S4000	*Set maximum spindle speed, spindle speed*
T0100M8	*Tool change, coolant on*
G00X5.Z2.T0101	*Rapid to tool change position, tool offset*
X3.5Z.100	*Rapid to clearance plane for first rough cut*
G01Z-3.95F.015	*First rough cut*
X3.48Z-3.93	*Retract tool at 45°*
G00Z.100	*Rapid to clearance plane*
X3.050	*Position for second rough cut*
G01Z-3.95F.015	*Second rough cut*
X3.030Z-3.93	*Retract tool at 45°*
G00Z.100	*Rapid to clearance plane*
X2.425	*Position for third rough cut*
G01Z-2.45	*Third rough cut*
X2.405Z-2.43	*Retract tool at 45°*
G00X5.Z2.T0100M9	*Rapid to tool change position, cancel tool offset, coolant off*
M5	*Spindle off*
M1	*Optional stop*
G96S500M4	*Constant surface speed control, surface speed value, spindle direction (CCW)*
G50S4000	*Set maximum spindle speed, spindle speed*
T0200M8	*Tool change, coolant on*
G00X2.5Z0T0202	*Rapid to position for facing, tool 2 offset 2*
G01X0	*Finish face workpiece*
Z.05	*Retract in Z axis*
G00X2.430	*Position to cut chamfer*
G01X2.37Z-.08F.007	*Cut chamfer*
Z-2.45	*Finish turn thread diameter (leaves .050)*
X3.0	*Face shoulder*
Z-3.850	*Finish turn 3 diameter*
G02X3.3Z4.0R.15	*Form .15 radius*

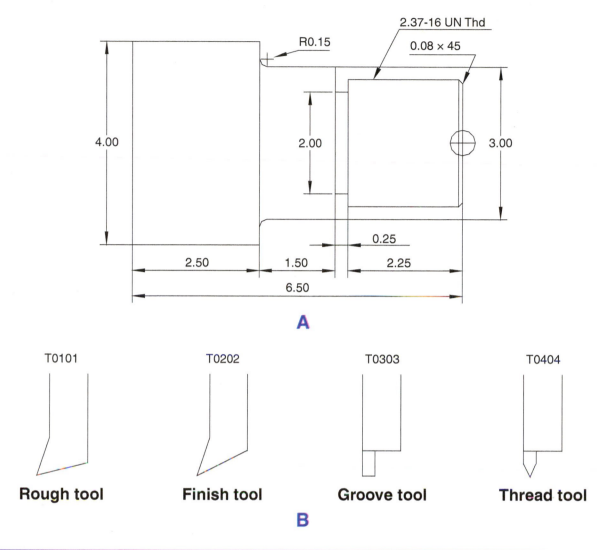

A

T0101	T0202	
T0303	T0404	

Rough tool   **Finish tool**   **Groove tool**   **Thread tool**

B

	T0101	T0202	T0303	T0404
Depth of Cut	0.250	0.010	0.200	4 passes
Feed Rate	0.015	0.007	0.004	Lead 0.0625
Spindle RPM				500
Cutting Speed	500 sfpm			

C

Figure 13-5. Sample program O2222. A— Part print. B—Tools used in program O2222. C—Cutting data used in program O2222 for turning, grooving, and threading operations.

G01X4.1	*Face shoulder*
G00X5.Z2.T0200M9	*Rapid to tool change position, cancel tool offset, coolant off*
M5	*Spindle off*
M1	*Optional stop*
G96S500M4	*Constant surface speed control, surface speed value, spindle direction (CCW)*
G50S4000	*Set maximum spindle speed, spindle speed*
T0300M8	*Tool change, coolant on*
G00X3.050Z-2.5T0303	*Rapid to position for grooving (undercut) tool 3 offset 3*
G01X2.00F.004	*Feed to groove depth*
G04U.5	*Dwell at bottom of groove for .5 seconds*
G01X3.05F.020	*Retract from groove*
G00X5.Z2.T0300M9	*Rapid to tool change position, cancel tool offset, coolant off*
M5	*Spindle off*
M1	*Optional stop*
G96S500M4	*Constant surface speed control, surface speed value, spindle direction (CCW)*
G50S4000	*Set maximum spindle speed, spindle speed*
T0400M8	*Tool change, coolant on*
G97S500	*Cancel CCS (constant cutting speed), 500 rpm*
G00X2.36Z.1	*Rapid to start position of threading*
G33Z-2.37F.0625	*First threading pass*
G00X2.5	*Retract threading tool*
Z.1	*Move to start plane*
X2.34	*Rapid to second start position*
G33Z-2.37	*Second threading pass*
G00X2.5	*Retract threading tool*
Z.1	*Move to start plane*
X2.324	*Rapid to third start position*
G33Z-2.37	*Third threading pass*
G00X2.5	*Retract threading tool*
Z.1	*Move to start plane*
X2.32	*Rapid to fourth start position*
G33Z-2.37	*Fourth threading pass*
G00X2.5	*Retract threading tool*
G00X5.Z2.T0400M9	*Rapid to tool change position, cancel tool offset, coolant off*
G28U0W0	*Return to home position*
M30	*Program end*

Turning—Example 3

Refer to **Figure 13-6** as you study this sample program. This example gives three components necessary in a programming process: a setup sheet, print, and tool list. Calculate the necessary tool positions needed for part programming from the part print.

O9239	*Program number*
G20G99G40	*Inch mode, feed per revolution, tool nose radius compensation cancel*
G96S300M4	*Constant surface speed control, surface speed value, spindle direction (CCW)*
T0100M8 (Facing MCFNL16-4)	*Tool change, coolant on*
G00X3.0Z2.T0101	*Rapid to tool change position, tool offset*
X2.7Z.125	*Rapid to clearance plane for first facing cut*
G01X-.0156F.006	*First facing cut*
G00Z.225	*Rapid to Z clearance plane*
X2.7	*Rapid to X clearance plane*
Z0	*Rapid to position for second facing cut*
G01X-.0156	*Final facing cut*
G00X3.0Z2.T0100M9	*Rapid to tool change position, cancel tool offset, coolant off*
M5	*Spindle off*
M1	*Optional stop*
G96S400M4	*Constant surface speed control, surface speed value, spindle direction (CCW)*
T0200M8 (Turning MTJNL16-4)	*Tool change, coolant on*
G00X2.35Z.1T0202	*Rapid to Z plane and first position for turning*
G01Z-2.2F.006	*Turn 2.35 diameter, 2.2 length*
X2.7	*Face shoulder to X clearance plane*
G00Z.1	*Rapid to Z clearance plane*
X2.25	*Position for second turning pass*
G01Z-2.2	*Turn 2.25 diameter, 2.2 length*
X2.7	*Face shoulder to X clearance plane*
G00Z.1	*Rapid to Z clearance plane*
X2.1	*Position for third turning pass*
G01Z-2.2	*Turn 2.1 diameter, 2.2 length*
X2.7	*Face shoulder to X clearance plane*
G00Z.1	*Rapid to Z clearance plane*
X2.0	*Position for fourth turning pass*
G01Z-2.2	*Turn 2.0 diameter, 2.2 length*
X2.7	*Face shoulder to X clearance plane*
G00X3.0Z2.T0200M9	*Rapid to tool change position, cancel tool offset, coolant off*
M5	*Spindle off*

Setup Sheet

Part Name: SBUSH2
Part Number: 09239
Program Number: 09239

Material: C1020
Material Size: 2.5 DIA. - 3.5 LG.

Directions: Butt workpiece against chuck face. Set part zero according to sketch.

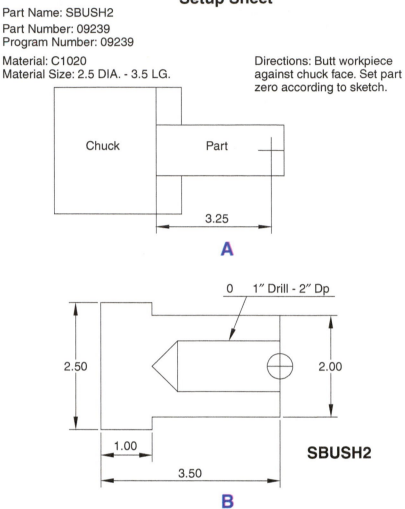

A

B

SBUSH2

CNC Tool List and Operation						
Part Name		**Part No.** O9239		**Operator No.** 1		**Unit No.**
SBUSH2		**Program No.** O9239		**Tool Library**		**Machine** Puma6HS
Tool Seq. No.	**Tool No.**	**Turret No.**	**Tool Description and Operation**	**Tool Offset**	**Tool Offset**	**Radius Comp.**
1	0101	1	Facing – MCFNL16-4			
2	0202	2	Turning – MTJNL16-4			
3	0303	3	Centerdrill			
4	0404	4	23/32 Drill			
5	0505	5	1″ Drill			

C

Figure 13-6. Sample program O9239 (Part SBUSH2). A—Setup sheet. B—Part print. C—Tool list.

M1	*Optional stop*
G97S3000M3	*Cancel CSS, 3000 rpm, spindle direction (CW)*
T0300M8 (Centerdrill)	*Tool change, coolant on*
G00X0Z.1T0303	*Rapid to Z plane and position for drilling, tool offset*
G01Z-.375F.005	*Centerdrill hole .375 deep, feed rate .005 ipr*
Z.1F.050	*Retract centerdrill at .050 ipr feed*
G00X3.0Z2.T0300M9	*Rapid to tool change position, cancel tool offset, coolant off*
M5	*Spindle off*
M1	*Optional stop*
S500M3	*500 rpm, spindle direction (CW)*
T0400M8 (23/32 Drill)	*Tool change, coolant on*
G00X0Z.1T0404	*Rapid to Z plane and position for drilling, tool offset*
G01Z-2.25F.006	*Drill hole, feed rate .006 ipr*
Z.1F.050	*Retract drill at .050 ipr feed*
G00X3.0Z2.T0400M9	*Rapid to tool change position, cancel tool offset, coolant off*
M5	*Spindle off*
M1	*Optional stop*
S300M3	*300 rpm, spindle direction (CW)*
T0500M8	*Tool change, coolant on*
G00X0Z.1T0505	*Rapid to Z plane and position for drilling, tool offset*
G01Z-2.0F.010	*Drill hole, feed rate .010 ipr*
Z.1F.050	*Retract drill at .050 ipr feed*
G00X3.0Z2.T0500M9	*Rapid to tool change position, cancel tool offset, coolant off*
G28U0W0	*Return to home position*
M30	*Program end*

Summary

The programming process should involve the steps of preparing a CNC coordinate drawing, listing the operations sequence, preparing a tool list, preparing a setup sheet, preparing a program manuscript, loading the program, verifying the program, running the program, and correcting and rerunning the program if necessary.

Chapter Review

Answer the following questions. Write your answers on a separate sheet of paper.

1. An MI sheet is sometimes called a(n) _____ sheet.
2. List three elements that could be found on a MI sheet.
3. List five items that could be found on a tool list.
4. Verifying a program by running the program without the part in the machine is called _____.
5. Two types of programming can be used: diameter programming and _____ programming.

Activities

Write two programs using the format given in this chapter. Use the provided print, tool list, and setup sheet for each program.

Program 1

Setup Sheet

Part Name: SBUSH3

Part Number: O9240
Program Number: O9240

Material: C1020
Material Size: Use part O9239

Directions: Butt workpiece against chuck face. Set Part Zero according to sketch.

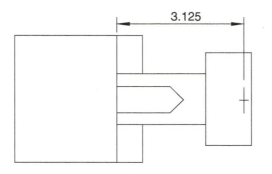

Setup Sheet

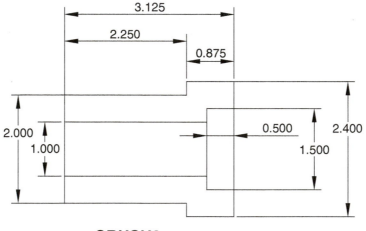

SBUSH3

Print

CNC Tool List and Operation

Part Name SBUSH3		Part No. O9240		Operator No. 1		Unit No.	
		Program No. O9240		Tool Library		Machine Puma6HS	
Tool Seq. No.	Tool No.	Turret No.	Tool Description and Operation	Tool Offset	Tool Offset	Radius Comp.	
1	0101	1	TN & FC – MTJNL16-4				
2	0303	3	Centerdrill				
3	0808	8	23/32 Drill				
4	1414	14	1″ Drill				
5	0505	5	Boring Bar				

Tool List

Program 2

Setup Sheet

Part Name: SBUSH3
Part Number: O9241
Program Number: O9241

Material: C1020
Material Size: Use part O9240

Directions: Butt workpiece against bottom
of chuck jaws. Grip workpiece by counterbore.

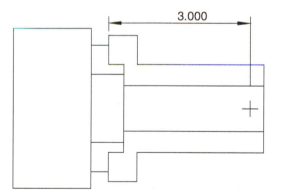

3.000

Setup Sheet

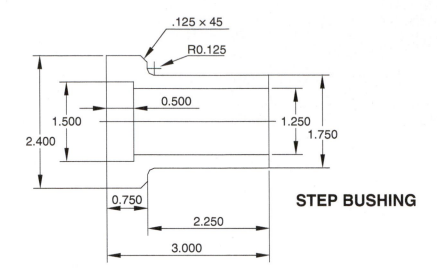

.125 × 45

R0.125

0.500

1.500

2.400

1.250

1.750

STEP BUSHING

0.750

2.250

3.000

Print

CNC Tool List and Operation						
Part Name		**Part No.** O9241		**Operator No.** 1		**Unit No.**
Step Bushing		**Program No.** O9241		**Tool Library**		**Machine** Puma6HS
Tool Seq. No.	**Tool No.**	**Turret No.**	**Tool Description and Operation**	**Tool Offset**	**Tool Offset**	**Radius Comp.**
1	0101	1	TN & FC – DCLNL-124A			
2	0202	2	TN & FC – DDJNL-124A			
3	0707	7	Boring Bar – BL8206			

Tool List

Chapter 14
Programming Codes for Turning Centers

Objectives

Information in this chapter will enable you to:

- Define various G-codes used in programs.
- Define various M-codes used in programs.
- Describe the use of offsets in a program.

Technical Terms

dwell command	machine home	tool nose radius (TNR)
first zero	preparatory codes	compensation
G-codes	second zero	wear offset

G-Codes

G-codes are commonly called *preparatory codes* because they determine the conditions under which the machine will function. These codes direct the equipment through various machining operations in addition to specifying axis moves. Refer to Chapter 8 for an explanation of the words X, Z, U, and W. **Figure 14-1** shows a partial listing of G-codes commonly used in turning center programs. Refer to the Appendix for a complete table of turning center G-codes.

G00—Rapid Positioning

This command moves a tool at a rapid traverse rate to a point (X, Z) in the work coordinate system. See **Figure 14-2**. U, W is an incremental movement.

Turning Center G Codes

Word	Group	Function
G00	1	Rapid travel positioning
G01	1	Linear interpolation
G02	1	Circular interpolation (CW)
G03	1	Circular interpolation (CCW)
G04	0	Dwell
G20	6	Inch data input
G21	6	Metric data input
G27	0	Reference point return check
G28	0	Return to reference point
G29	0	Return from reference point
G30	0	Return to second reference point
G32	1	Straight and taper cutting
G40	7	Tool nose radius compensation cancel
G41	7	Tool nose radius compensation left
G42	7	Tool nose radius compensation right
G50	0	Programming of absolute zero point
G96	2	Constant surface speed control
G97	2	Constant surface speed control cancel
G98	5	Feed per minute
G99	5	Feed per revolution

Figure 14-1. Functions of various G-codes used with turning centers.

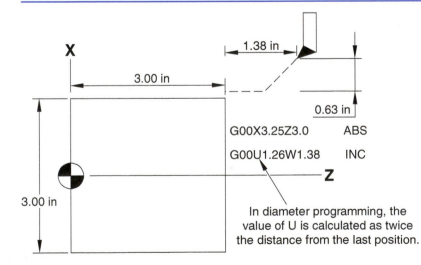

G00X3.25Z3.0   ABS

G00U1.26W1.38   INC

In diameter programming, the value of U is calculated as twice the distance from the last position.

Figure 14-2. An example of rapid tool path movement to the end of a workpiece.

G01—Linear Interpolation

Linear interpolation moves a tool in a straight path to the (X, Z) position in the work coordinate system, or from its present position to the position specified by the incremental values (U, W). See **Figure 14-3**. This command requires that a feed rate (F) be stated.

G02—Circular Interpolation (CW) and G03—Circular Interpolation (CCW)

A CNC machine control unit has the capability of moving a tool in a circular arc. To accomplish this move; three pieces of information must be specified in the command:

- The endpoint of the move

- The location of the arc center (specified as the incremental distance from the start of the arc to the center of the arc)

- The direction of rotation of the arc

The endpoint of the arc is specified in a similar manner as a straight-line move. The arc center is specified as if moving from the start point to the arc center in an incremental manner. Since *X* and *Z* are used to specify the endpoint of the arc, the letter *I* and *K* are used to specify the arc center distance, substituting for the letters *X* and *Z*. *I* represents *X* and *K* represents *Z*.

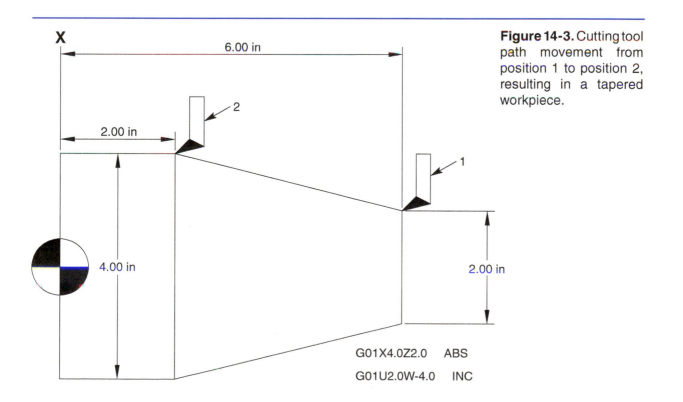

Figure 14-3. Cutting tool path movement from position 1 to position 2, resulting in a tapered workpiece.

G01X4.0Z2.0   ABS

G01U2.0W-4.0   INC

Direction of the circular move may be clockwise (G02) or counterclockwise (G03). The universal method of specifying a circular move uses *X* and *Z*, *I* and *K*, and either G02 or G03. Many controllers can substitute the letter *R* (radius) to designate the radius of the arc, thereby eliminating the need for the letters *I* and *K*. **Figure 14-4** shows a partial program that uses both methods of specifying a circular motion. Movement from position 2 to position 3 uses the *I, K* method of specifying an arc. Movement from position 4 to position 5 uses the radius method of specifying a circular motion.

Note

Further clarification or explanation regarding G02 and G03 moves is provided in Chapter 8, which applies these codes to machining centers that function in the XYZ coordinate system.

G04—Dwell

A *dwell command* will stop all *X* and *Z* movement. It is primarily used to break metal chips at the end of a machining operation, such as necking, grooving, boring, or turning to a shoulder. This allows cleanup on the workpiece for surface finish as well as size control. Dwell can also be used to allow time for a spindle to reach a programmed speed when there is a vast change in rpm.

The dwell command can be specified with the X-, U-, or P-words. Dwell is based on seconds, with the maximum command time at 99999.999 seconds. For example, a dwell for 1.25 seconds can be specified as G04 X1.25, G04 U1.25, or G04 P1250. A decimal point is not used with the P-word.

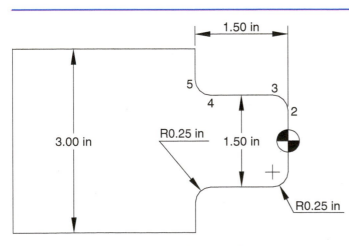

Figure 14-4. Commands resulting in a tool path consisting of two circular moves and one linear (straight-line) move.

G03X1.5Z-.25I0K-.25   Movement from 2 to 3
G01Z-1.25   Movement from 3 to 4
G02X2.0Z-1.5R.25   Movement from 4 to 5

G27, G28, G29, G30—Zero Return Functions

Many CNC lathes have two machine zero positions called first zero and second zero. The *first zero* position is located to the extreme right-hand corner of the machine. The *second zero* position is located a fixed distance from the first zero and is established by setting parameters. A program is started from one of these zero points. If the program is interrupted in the middle of a auto run, the operator can locate the start point of the program by returning to one of these zero points. A command that sends a machine slide to the first zero causes the slide to travel in rapid speed until it is approximately 1/4″ from the zero position. Then, the machine parameters guide the ball screws, regulating how much of a revolution is needed to reach the zero (*machine home*) position.

The machine operator positions the turret slide at the machine home position and starts the program. At the end of the program, the turret should be returned to the machine home position.

The turret slide can be positioned at the second zero location using a G-code. The second zero position is usually used as a tool change position. The G27 (first zero return check) is usually not used in programs and is only used to check if home position is reached. G28 (automatic first zero return) is frequently used to return the turret slide to machine home through an intermediate point specified by addresses X and Z. See **Figure 14-5**. G29 (automatic return for first or second zero) provides positioning to a commanded point through an intermediate point. The intermediate point must be already set by a previous G28 command.

The G28 command is a rapid traverse move and can be used when there is a device in the path, such as a tailstock. See **Figure 14-6**. If a path is specified to return to machine zero using G28 to clear the tailstock, the same path could be used to position the tool for the next move back to the part.

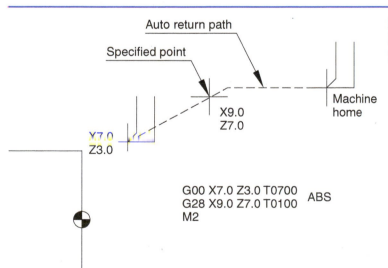

Figure 14-5. Tool path to machine home position through a specified intermediate point.

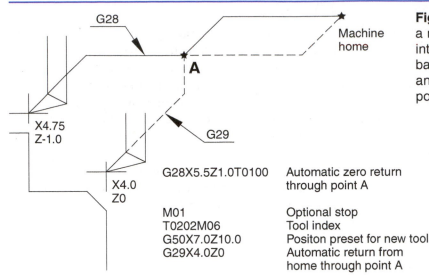

Figure 14-6. Tool paths showing a return to machine home through intermediate point (G28) and going back to the workpiece through an intermediate point to a new position (G29).

G28X5.5Z1.0T0100	Automatic zero return through point A
M01	Optional stop
T0202M06	Tool index
G50X7.0Z10.0	Positon preset for new tool
G29X4.0Z0	Automatic return from home through point A

The G30 command (automatic second zero return) permits positioning to the second zero point through a point defined by the addresses X and Z, **Figure 14-7**. The second zero point is set by entering parameters as a distance from the machine zero point. The command is the same as the first zero return (G28), except that the tool returns to the second zero (reference) point and not the machine home. The G30 command is usually used when the first zero point (home) is not going to be used as the program starting point, because of the additional time needed to travel to home.

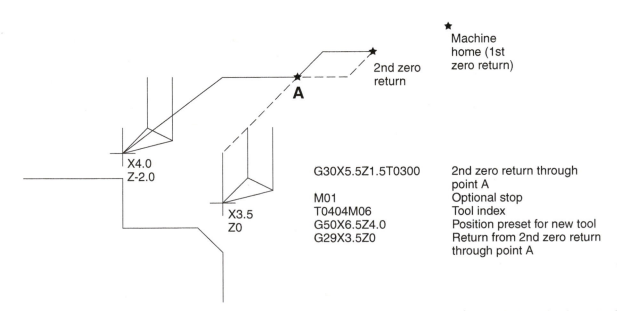

G30X5.5Z1.5T0300	2nd zero return through point A
M01	Optional stop
T0404M06	Tool index
G50X6.5Z4.0	Position preset for new tool
G29X3.5Z0	Return from 2nd zero return through point A

Figure 14-7. Tool paths showing a return to second zero return through an intermediate point for a tool change (G30) and returning to a new position through the intermediate point (G29).

G33—Thread Cutting

A G33 command is used for both straight and taper threads. Threads are started at a fixed point and the tool path remains unchanged until all thread passes are complete. The spindle speed must remain constant throughout the entire rough and finishing sequence of threading. Before programming the cutting of a thread, determine how many passes will be made to cut the thread and how deep the tool will be fed into the work for each pass. Technical tables in cutting tool catalogs provide that type of information. The whole depth of the thread can be obtained from reference tables found in references such as *Machinery's Handbook*. The number of passes to be made may vary, depending on the class fit of the thread and the required finish. Fine threads have a shallow thread depth and require fewer passes than coarse threads, which have a greater thread depth value.

Thread lead is determined by the formula $P = 1/N$, where P is pitch and N is the number of threads per inch. The letter *F* is usually used to specify feed. In some cases, however, the letter *E* is substituted for feed.

Note

Some controllers use G32 as the code for threading.

The following example is a partial program showing the threading commands for a 2"-20 thread. See **Figure 14-8**. A table from a manufacturer's catalog and technical guide can be used to determine the number of passes for a thread this size being cut in carbon steel. As shown in **Figure 14-9**, this thread calls for six passes with the depths shown, for a total depth of .033".

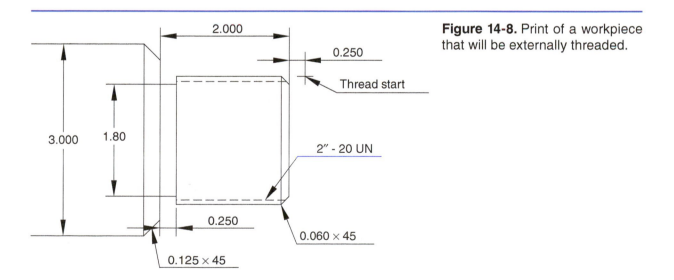

Figure 14-8. Print of a workpiece that will be externally threaded.

Depth per Pass					
First pass	**Second pass**	**Third pass**	**Fourth pass**	**Fifth pass**	**Sixth pass**
0.008	0.007	0.006	0.005	0.004	0.003

Figure 14-9. A reduced depth of cut is performed for each successive thread pass.

T0400M8	
G97S400	
G00X1.984Z.250	*Starting point of thread/first depth*
G33Z-1.940F.05	*First pass*
G00X2.050	*Retract from thread*
Z.250	
X1.970	*Second depth*
G33Z-1.94F.05	*Second pass*
G00X2.05	*Retract from thread*
Z.250	
X1.958	*Third depth*
G33Z-1.94F.05	*Third pass*
G00X2.05	*Retract from thread*
Z.250	
X1.948	*Fourth depth*
G33Z-1.94F.05	*Fourth pass*
G00X2.050	*Retract from thread*
Z.250	
X1.940	*Fifth depth*
G33Z-1.94F.05	*Fifth pass*
G00X2.050	*Retract from thread*
Z.250	
X1.934	*Sixth depth*
G33Z-1.94F.05	*Sixth pass*
G00X2.050	*Retract from thread*
X5.0Z2.0T0400M9	*Rapid to tool change position/cancel tool offset/coolant off*

Tool Offset

Tool offsets allow the machine operator to manually change the programmed tool position. By performing an offset, the operator can make minor tool position adjustments and compensate for tool tip wear and tool deflection. Offsets are made active by specifying the offset number in the program using the last two digits of the T-word. For example, in the T-word T0203, the last two digits (03) represent the offset number.

Wear offset is used to adjust part size. The machine operator enters offset values into the offset file during machine setup and actual running of the job. Adjustments usually have to be made when running many parts, because of tool wear. *Tool nose radius compensation (TNR)* is used to compensate for varying tool tip radii values. The operator enters the value of the tool nose radius and the number for the shape of the tool in the offset registers. Geometry offsets are used to compensate for tool size differences. Tool offset must be made active prior to starting radius compensation.

Tools are qualified to length using a tolerance of ±0.003". When a cutting insert is changed because of excessive wear, breakage, or chipping, the location of the new insert can vary within ±0.003". If a tolerance of ±0.001" has to be held, the tool tip position has to be corrected at every insert change. This requires changing the G50 or G92 values, which are discussed later in this chapter.

Rather than revising the program, the tool point can be moved by a specific amount by inserting a value into the tool offset file. In a program, tool offset is a two-digit number assigned to a specific tool. For example, T0304 means tool number 3 has an offset file value of 4. The actual offset amount, such as 0.0025", is not specified in the program. It is inserted into the number 4 offset file. Offset values assigned to the tool number are added algebraically (plus or minus).

Here is an example of applying an offset value: Assume, in a two-tool program, that tool 2 is 0.010" shorter than tool 1. A value of .010 is placed in the X offset file for tool 2, thereby making Tool 2 equal in length to Tool 1. If a tool is narrower or wider than another tool, an offset value equivalent to the difference in the tools is placed in that tool's Z offset file.

When the program is initially written, the programmer does not know what the tool offset value will be. The value is determined during the machine setup, with the offset value manually input into the memory file. If an insert is changed during the execution of a job, the tool offset value must also be changed if the size tolerance this insert is to hold is less than the tolerance of the insert change.

Turning center tool offsets are always in pairs, since cutting is performed in two axes (X and Z). Tolerances must be held in both directions. Each offset number has two values, one for X and the other for Z. Values can be entered in either or both of these axes. When this offset number appears in the program, it will alter the cutting tool position by the offset value amount found in the offset registers.

Remember, tool offsets are associated with the tool, not the program. Therefore, tool offsets will remain with the tool and the values will remain, regardless of the program number. No matter how many jobs use this tool, the offset value will remain in effect as long as that tool is not removed from the turret. If machining variables change, it may be necessary to change offset values when new jobs are started.

G41, G42—Tool Nose Radius Compensation

The programming tool point, until now, has been referred to as Point P, which is an imaginary point. See **Figure 14-10**. Since the tool tip is not a sharp point, but a radius, *tool nose radius (TNR) compensation* is used to correctly size and shape the workpiece when angles and arcs need to be cut on a part. Tool nose radius values commonly used are 0.0156", 0.0312", 0.0469", and 0.0625".

The tool nose should be thought of as a circle moving along the workpiece. The circle remains tangent to the surface of the workpiece as it travels the cutting path. If the nose radius loses its tangency to the part at any given time, the workpiece shape will be inaccurately altered.

When cutting parallel to the Z axis, actual cutting occurs at Point B. Because line PB is perpendicular to the X axis, Point P may be used for programming. However, Z distances might not be accurate, since P is a theoretical point and not an actual point. If all workpiece surfaces are parallel to both the X and Z axes, then using Point P instead of A or B is allowed. See **Figure 14-11**. In this case, programming the intersection of the surfaces as the program points will produce the required shape.

If workpiece surfaces are angular and circular to the axes, then using Point P for programming results in a shape error. See **Figure 14-12**. The desired arc path is different from the actual path. The error is caused when the tool nose radius loses its tangency to the cutting surface. The tool nose was programmed to the intersection of the surfaces. This is why tool nose radius compensation is beneficial. Otherwise, tool position values have to be recalculated by the programmer and entered into the program.

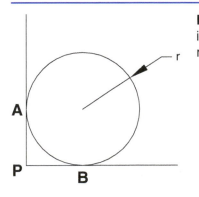

Figure 14-10. The circle represents the tool radius, with A and B indicating the material contact points when facing and turning. Point P represents the imaginary sharp point of the tool.

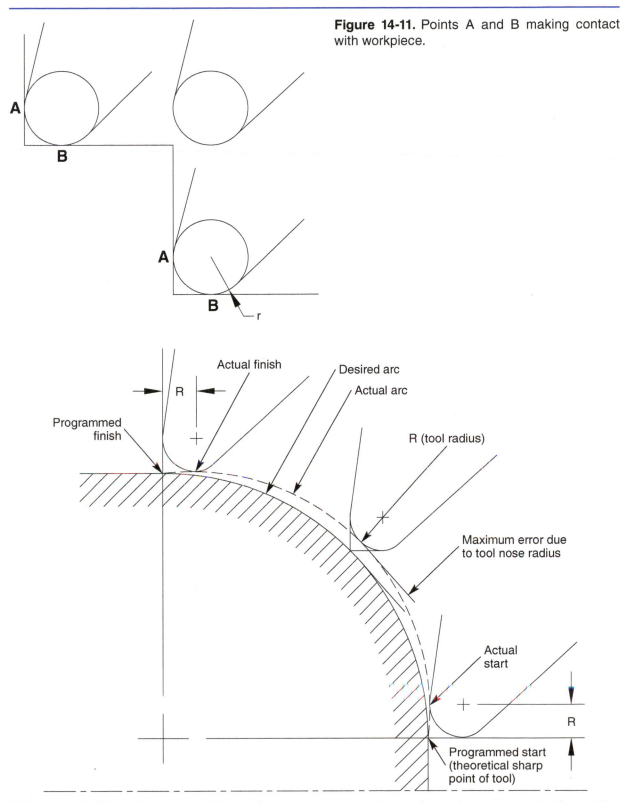

Figure 14-11. Points A and B making contact with workpiece.

Figure 14-12. The effect on workpiece shape when the tool nose loses its tangency to the cutting surface.

Certain information must be provided and certain actions taken when employing tool nose radius compensation:

- Direction of the standard tool tip
- Stated direction of the cutter path (G41/G42)
- Tool compensation vector
- Setting of the tool tip and nose radius in the offsets table
- Activating TNR compensation in the start-up block
- Canceling TNR compensation in the cancel block

Standard Tool Tip Direction

There are nine possible positions for a tool in relation to the workpiece when using nose radius compensation. These positions are relative to a full (360°) circle. See **Figure 14-13**.

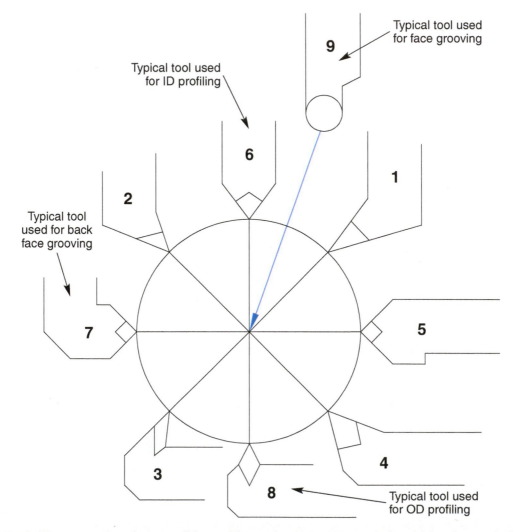

Figure 14-13. These are the nine possible positions of a tool with its relationship to the workpiece.

Rules for Using TNR Compensation

The following is a list of rules for using tool nose radius compensation:

- A G-code command initializing TNR compensation should be given in a separate block of information and should be in effect before the tool starts cutting.

- The G-code (either G41 or G42) must be followed by an axis linear command that initiates the offset move to take place; for example, G01X3.0Z2.0.

- Motion distance in the G40, G41, and G42 commands should be greater than the amount of the tool nose radius value.

- Canceling of TNR compensation should be done with a G40 command. This command is stated after the tool exits the workpiece to avoid overcutting or undercutting due to the offset move that occurs when compensation is removed.

- G02 and G03 cannot initiate TNR compensation.

- G40 cancels the TNR compensation.

- G41 is TNR compensation left.

- G42 is TNR compensation right.

- When viewing travel of the tool from behind as it moves away from the operator, if the tool is to the right of the surface being cut, it is a G42 condition. If the tool is to the left of the surface being cut, it is a G41 condition. See **Figure 14-14**.

Tool Compensation Vector

When the G41or G42 command is applied to an offset, an offset vector (path) is created. This offset vector has a distance equal to the nose radius as specified in the control offset file. This vector is created by the control unit when G41 or G42 is specified. It is used to calculate a tool path that is offset from the programmed path by an amount equal to the tool nose radius. After the initial vector takes effect, the machine controller reads ahead in the program and continues calculating this vector block by block based on tool movement. **Figure 14-15** shows the startup vector move for G41 and G42 commands.

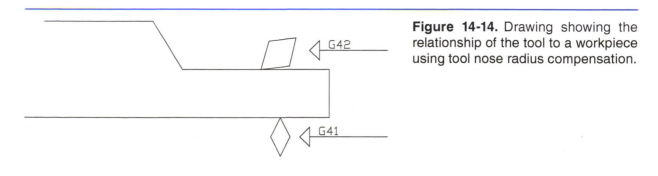

Figure 14-14. Drawing showing the relationship of the tool to a workpiece using tool nose radius compensation.

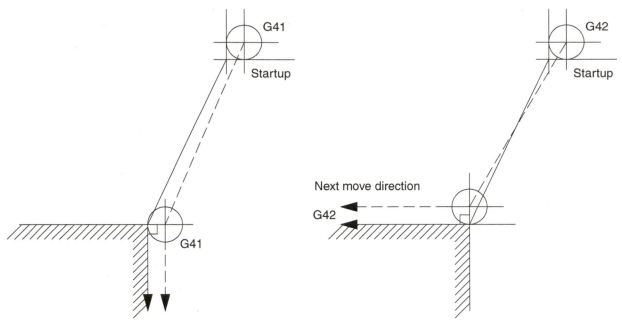

Figure 14-15. Startup vector moves for G41 and G42 (offset vector) commands.

TNR Compensation—Example 1

Refer to **Figure 14-16** as you study this sample program. This example shows a portion of a program using TNR compensation to finish turn and face a workpiece. Note that the TNR compensation commands are modal.

N10G20G99G40	*Inch mode, fpr, TNR compensation cancel*
N20G95S300M3	*CCS, surface speed value, spindle CW*
N30G50S3000	*Set max spindle speed, spindle speed*
N40T0100M8	*Tool change, coolant on*
N50G00X8.Z4.T0101 (Pos 1)	*Rapid to position 1*
N60G41	*TNR compensation left*
N70G01X4.8Z0F.005 (Pos 2)	*Linear move to position 2, .005 ipr*
N80X2.75 (Pos 3)	*Feed to position 3*
N90G40	*Cancel TNR compensation*
N100G01Z.25 (Pos 4)	*Feed to position 4*
N110G42	*TNR compensation right*
N120G01X3.0 (Pos 5)	*Feed to position 5*
N130X4.0Z-.25 (Pos 6)	*Feed to position 6*
N140Z-1.25 (Pos 7)	*Feed to position 7*
N150X5.0 (Pos 8)	*Feed to position 8*
N160X6.0Z3.0 (Pos 9)	*Feed to position 9*
N170X7.0 (Pos 10)	*Feed to position 10*

N180G40   *Cancel TNR compensation*
N190G00X8.0Z4.0T0100M9   *Rapid to tool change position, cancel tool offset, coolant off*
N200G28U0W0   *Return to home position*
N210M30   *Program end*

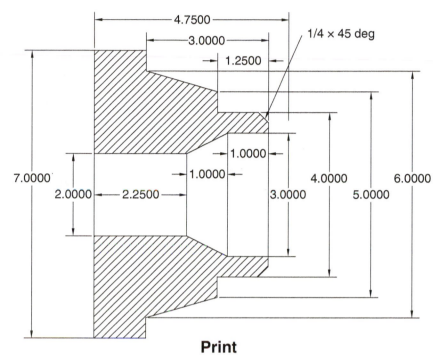

Print
A

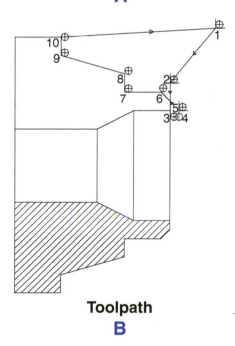

Toolpath
B

Figure 14-16. This is the print and tool path for Example 1.

G50—Coordinate System Designation

Before programming with respect to a part print, it is necessary to establish a coordinate system or set a coordinate zero (part or program zero) for each axis. This is necessary to dimension the cutting tool position in relation to the workpiece being machined. It is common to set the program zero at certain given points. See **Figure 14-17**. When part zero is located at the tailstock end of the workpiece, the cutting tool is referenced as G50X3.9084Z1.0438 (diameter programming) or G50X1.9542Z1.0438 (radius programming).

When part zero is located at the chuck end of the workpiece, the cutting tool is referenced as G50X3.9084Z7.7938 (diameter programming) or G50X1.9542Z7.7938 (radius programming). When this method is used (normally on older control models), each tool must be set to part zero. Either a tool change position or home position is used as reference. When the tool is finished cutting, it is sent back to the same coordinates used in the G50 command. The next tool used in the turret can be set to part zero. Some controls use the G92 code for this purpose because they employ a different G-code list.

Maximum Spindle Speed

When a G50 code is used with a spindle speed code, such as S350, the maximum limit for spindle rpm is set. The G50 code is used in combination with a G96 code, which invokes a constant cutting speed. The G96 code is discussed in the next section of this chapter. In a program, the G50 command combined with the G96 command appears as the following:

G50S2000
G96S300

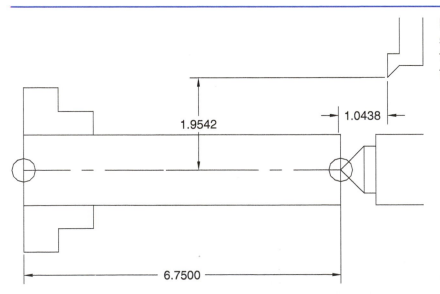

1.9542

1.0438

6.7500

Figure 14-17. This drawing shows the relationship of the tool tip to part zero at both the tailstock and chuck ends.

This means that a constant surface speed of 300 surface feet per minute should be maintained by the spindle at all times, but the maximum rpm attained should not exceed 2000.

G96, G97—Constant Surface Speed and Cancel CSS

The constant surface speed (CSS) mode is used when the spindle must maintain a constant surface speed to obtain a uniform finish. As the tool moves away or toward the workpiece centerline, the tool rpm increases or decreases to maintain the specified surface speed given by the G96 command. As the tool moves away from the workpiece center (X+ direction), rpm decreases. As the tool moves toward the workpiece center (X– direction), rpm increases. This continues until the maximum rpm specified in the G50 command is reached.

If a taper, arc, or face is finished in the direct rpm mode without the G96 function, the surface finish is not evenly smooth because the changing diameter is cut at the same rpm. Also, if the spindle speed is changed in steps while in the cut, it will not result in uniform improvement of the finish. Because the feed of the axis, or axes, is stopped momentarily at the time of the speed change, a mark will appear on the finished surface.

G96 will place the control in the constant cutting speed mode, while G97 cancels cutting speed and uses direct rpm.

Program without G96

```
N90G00X5.Z1.0
N100G97S300
N110G01X2.75F.007
N120Z1.3
N130G97S550
N140X2.25
N150Z1.6
N160G97S800
N170X0
```

Program with G96

```
N90G00X5.Z1
N100G96S350
N110G01X2.75F.007
N120Z1.3
N130Z2.4
N140Z1.5
N150X0
```

Depending on the machine controller, the maximum rpm reached in constant surface speed can be limited by using either a G50 S*nnn* command or a G92 S*nnn* command. These commands inform the computer not to exceed the rpm value beyond the S (spindle speed) value. For example, G50S800 indicates a maximum speed of 800 rpm and a G92S900 indicates a maximum speed of 900 rpm. Following is a partial program using G92:

N80G00X5.0Z1.0

N90G50S1200

N100G96S250

N110G01X2.75F.007

N120Z1.3

N130Z2.4

N140Z1.5

N150X0

Summary

The word address code for rapid movement is G00. The word address code for linear interpolation is G01. A G02 command makes a CW arc movement. A G03 command makes a CCW arc movement. A G04 command causes a dwell.

G20 and G21 commands state inch and metric values. The G27 code checks to determine whether zero return has occurred. G28 returns the turret slide to machine home through an intermediate point. G29 moves the turret slide to a previously defined point (G28) through an intermediate point.

G30 positions the turret slide to a second zero point through an intermediate point. G33 is a command for cutting threads. The word address command for tool compensation left is G41. Tool compensation right uses a G42 command. The code G40 cancels tool compensation.

The G50 command is used to set part zero or specify a maximum spindle speed. G96 places the control in a constant cutting speed mode. G97 cancels constant speed.

Chapter Review

Answer the following questions. Write your answers on a separate sheet of paper.

1. What does the letter code *I* represent?
2. What does the letter code *K* represent?
3. What does *dwell* mean?
4. What words can be used to specify a specific time in a dwell command?
5. What command is used to return the turret slide to machine home?
6. Why would a G29 command be used?
7. What formula is used to determine thread pitch when thread cutting?
8. The last two digits in a T-word represent what value?

9. What are geometry offsets used for?
10. Workpiece size tolerances can be controlled using what type of offsets?
11. What type of tool uses an offset value of 2?
12. How many tool tip directions are available for use?
13. What two types of programming are used on machines?
14. What does the command G50S800 mean?
15. What command tells the controller to use feed per minute?

Activities

1. Write a circular arc command using R from the example shown. Movement is from A to B.

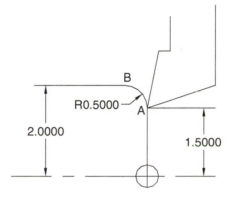

2. Write a circular arc command using IJK from the example shown. Movement is from A to B.

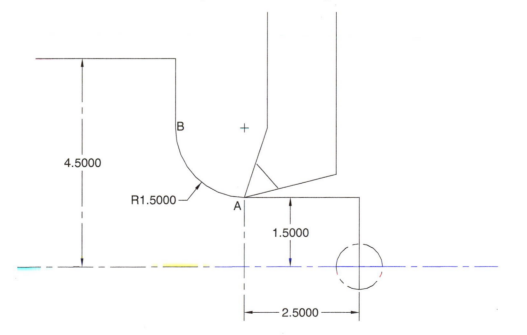

3. Write a dwell command for a 1.5 second dwell using X5.3 format and a 2 second dwell using P5.3 format.

4. Determine the G50 X and Z values for the example shown. For Tool A, find the values using diameter programming and radius programming. For Tool B, find the values using diameter programming and radius programming.

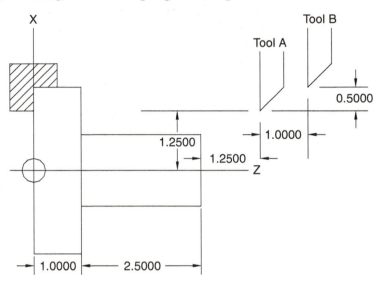

Chapter 15
Canned Cycles for Turning Centers

Objectives

Information in this chapter will enable you to:

- Recognize and define the various components of a turning center canned cycle.
- Write several canned cycles for a machine part.

Technical Terms

peck drilling
profiling cycle
spring pass

Canned Cycles

Canned cycles are used to greatly shorten the length of programs that require repetitive moves to perform certain cutting operations. The programmer defines the finished tool path shape and provides cutting parameters. The tool automatically removes a specific amount of material per pass until the finished shape is completed.

Seven canned cycles (sometimes referred to as multiple repetitive cycles by some controller manufacturers) will be defined and illustrated in this chapter. These canned cycles are G71, G70, G72, G73, G74, G75, and G76. There are various cycles used by other controllers; however, this chapter will cover cycles used on Fanuc controls. Simple canned cycles such as G90, G92, and G94 will also be discussed.

G71—Turning and Boring Cycle

The G71 cycle is used primarily for roughing the diameters of a workpiece, followed by a G70 cycle that finishes the diameters of the workpiece. However, the G71 cycle can be used alone to machine the workpiece if finish is not a major concern. The format of using a G71 cycle is as follows:

G71U*n*R*e*

G71P*ns*Q*nf*U*n*W*n*F*n*

Where:

- **G71.** Initiates the cutting cycle
- **U*n*.** On the first line, it specifies depth of cut per side for each cut
- **R*e*.** Escape (retract) amount
- **P*ns*.** Line number at the beginning of the cycle
- **Q*nf*.** Line number at the end of the cycle
- **U*n*.** On the second line, it specifies the direction and finish allowance along the X axis
- **W*n*.** Specifies direction and finish allowance along the Z axis
- **F*n*.** Feed rate

Rough machining using G71

The process of rough machining using G71 can be described in three steps:

1. The tool is sent to a starting position. For turning, the starting position is 0.100" from the end of the workpiece and flush with the outside diameter. For boring, the starting position is also 0.100" from the end of the workpiece and flush with the existing hole size.

2. Rough turning or boring begins using the information given in the G71 command. The tool rapids down in the X axis to position for a cut. It then feeds along the Z axis until the tool nears what would be a Z coordinate on the finish path (the control calculates the Z values in order to determine where to stop the tool to avoid cutting into the finish profile). The tool feeds off the cuts (approximately 0.010"), then rapids back to the Z plane (0.100" off end of workpiece). The cycle repeats itself, moving into the workpiece the amount of depth specified in the G71 command.

3. Cutting is performed along the defined finish profile, leaving stock for finishing. The cycle ends with a return to the initial starting position. See **Figure 15-1**.

Note

Cutting the finish profile is achieved with a G70 command.

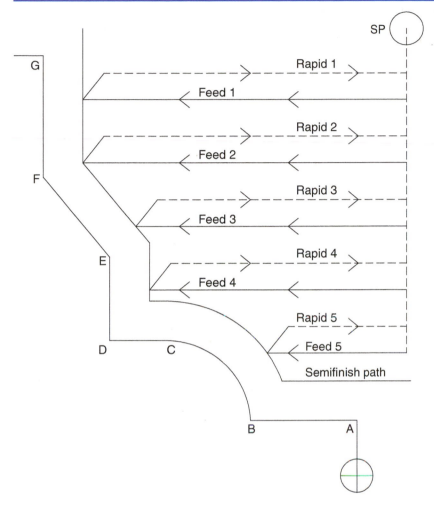

Figure 15-1. The tool path taken while performing a semi-finishing turning operation using a G71 canned cycle. Refer to Figure 15-3 for a print of the workpiece being machined.

G70—Finish Turning and Boring Cycle

The G70 cycle is used to finish machine the material left on the workpiece after the G71 roughing cycle. The format of using a G70 cycle is as follows:
G70P*ns*Q*nf*

Where:

- **P***ns*. The number at the beginning of the cycle

- **Q***nf*. The line number at the end of the cycle

The finishing cycle begins at the same starting point and ends at the point that was used for the roughing cycle. See **Figure 15-2**.

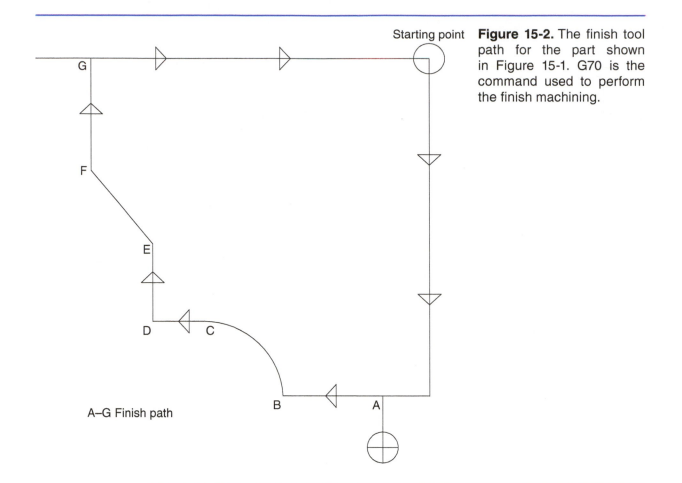

Starting point

Figure 15-2. The finish tool path for the part shown in Figure 15-1. G70 is the command used to perform the finish machining.

A–G Finish path

G71 and G70

This example covers G71 and G70 canned cycles. See **Figure 15-3.**

N10G20G99G40	*Inch mode, fpr, TNR compensation cancel*
N20G95S500M3	*CCS, surf. speed value, spindle CW*
N30G50S1000	*Set max. spindle speed, spindle speed*
N40T0100M8	*Tool change, coolant on*
N50G00X8.0Z4.0T0101	*Rapid to XZ location, apply offset to tool*
N60X7.0Z.100	*Rapid to XZ position*
N65G71U2500R.020	*Initiate cutting cycle, 0.250 doc, esc. amt.*
N70G71P80Q200U.040W.005F.012	*L80 cycle start, L200 cycle end, finish allowance, feed rate*
N80G00X2.0	*Lines 80–200 tool path*
N90G42	
N100G01Z-.5F.008	
N110X2.5Z-1.0	
N120Z-3.75	
N130X4.0	

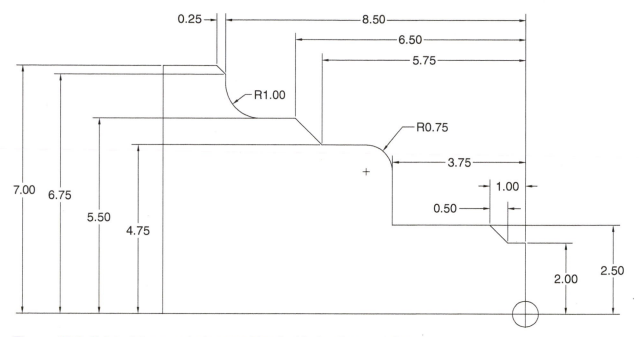

Figure 15-3. Print of the part being machined with the G71 and G70 cycles.

N140G03X4.75Z-4.5R.75
N150G01Z-5.75
N160X5.5Z-6.5
N170Z-7.5
N180G02X6.5Z-8.5R1.0
N190G01X6.75
N200X7.1Z-8.8

N210G00G41X3.5Z.020	*Rapid mode, TNR comp. left, position to XZ*
N220G01X0F.010	*Facing cut*
N230G40	*Cancel radius comp*
N240G00X7.0Z.1	*Rapid from work*
N250X8.0Z4.0T0100	*Rapid to tool change position, cancel tool offset*
N260M5	*Spindle stop*
N270M1	*Optional program stop*
N280G96S400M3	*CCS, 400 sfpm, CW spindle*
N290G50S3000	*Max. spindle speed set at 3000 rpm*
N300T0200M8	*Tool change, turn on coolant*
N310G00X2.5Z.1	*Rapid to XZ position*
N320G41	*TNR compensation left*
N330G01Z0F.05	*Position for finish face cut*
N340X0F.005	*Finish face end of workpiece*
N350G40	*Cancel TNR compensation*
N360G00X7.0Z.1	*Rapid to XZ position*

N370G70P80Q200	*Finish canned cycle, lines 80–200*
N380G40	*Cancel TNR compensation*
N390G00X8.0Z4.0T0200M9	*Rapid to tool change position, cancel tool offset, turn coolant off*
N400G28U0W0	*Return to tool home*
N410M30	*End program*

G72—Rough and Finish Facing Cycle

Using a facing cycle to machine short pieces is more efficient than using a turning cycle to remove the excess material. A facing cycle works well on large-diameter workpieces. The words used in a G72 command are the same as those used in a G71 command. Therefore, the format appears as:

G72W*n*R*e*

G72P*ns*Q*nf*U*n*W*n*F*n*

Where,

- **G72.** Initiates the cutting cycle
- **W*n*.** Specifies depth of cut on the first line
- **R*e*.** Escape (retract) amount
- **P*ns*.** Line number at the beginning of the cycle
- **Q*nf*.** Line number at the end of the cycle
- **U*n*.** Specifies the direction and finish allowance along the X axis on the second line
- **W*n*.** Specifies the direction and finish along the Z axis on the second line
- **F*n*.** Feed rate

The process of rough facing using G72 is essentially the same as the G71 roughing cycle. With G72, however, the cutting takes place along the X axis rather than the Z axis used with G71.

U and W values need the correct sign to show in which direction stock should be left. **Figure 15-4** shows two patterns representing U and W being used to define stock to be left on the workpiece while performing OD and ID facing operations. **Figure 15-5** shows the tool path that is taken when performing the G72 canned cycle.

G72 and G70

This example covers G72 and G70 canned cycles. See **Figure 15-6**.

N80G00X8.6Z.100	
N90G72W1500R.050	*Facing cycle, 150" doc, .050" retract amt.*
N100G72P110Q160U.015W.010F.015	*Cycle lines 110–160, 0.015" finish allowance on diameter (X), 0.010" finish allowance on faces (Z), 0.015 ipr feed rate*

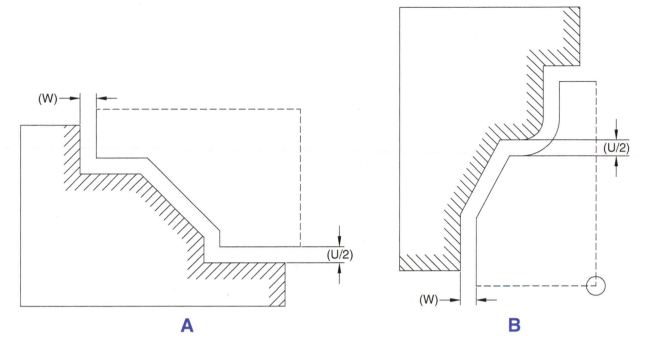

Figure 15-4. Facing operations. A—U and W being used to define stock to be left on the workpiece while performing an outside diameter (OD) facing operation. B—U and W being used to define stock to be left on workpiece while performing an inside diameter (ID) facing operation.

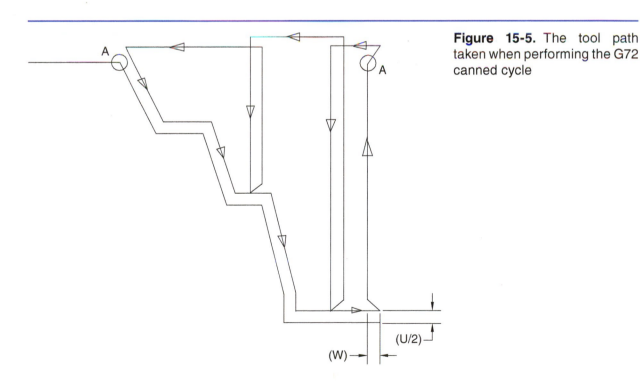

Figure 15-5. The tool path taken when performing the G72 canned cycle

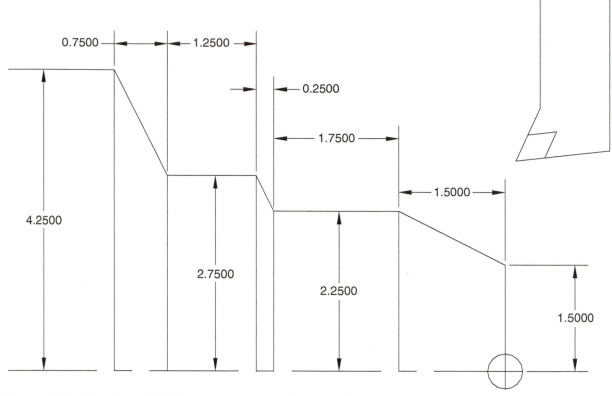

Figure 15-6. Rough and finish facing using the G72 and G70 canned cycles.

N110G00Z-5.6

N120G01X5.5Z-4.75 (Position 5)

N130Z-3.5 (Position 4)

N140X4.5Z-3.25 (Position 3)

N150Z-1.5 (Position 2)

N160X2.9Z.1

N170G70P110Q160   *Finish pass for canned cycle*

G73—Pattern Repeating Cycle

The G73 cycle allows the repeated cutting of a fixed pattern, with the pattern being shifted little by little. Some control manufacturers call this a *profiling cycle*. With this cutting cycle, it is possible to efficiently machine a workpiece with a rough shape made by a forging, casting, or rough machining operation. **Figure 15-7** shows the pattern of a G73 cycle.

The G73 cycle is quite similar to other cycles, such as turning and facing, except for some minor addition of words. The format of using a G73 cycle is as follows:

G73U*n*W*n*R*e*

G73P*ns*Q*nf*U*n*W*n*F*n*

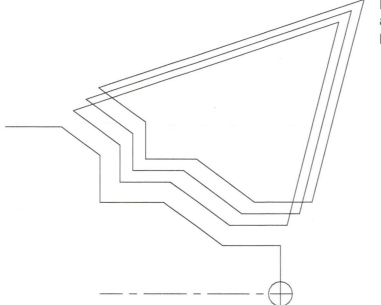

Figure 15-7. The G73 canned cycle allows repeated cutting of a fixed pattern.

Where,

- **G73.** Initiates the cutting cycle
- **U***n***.** On the first line, it specifies the relief distance and direction of the X move
- **W***n***.** On the first line it specifies the relief distance and direction of the Z move
- **R***e***.** Escape (retract) amount
- **P***ns***.** The number at the beginning of the cycle
- **Q***nf***.** The line number at the end of the cycle
- **U***n***.** On the second line, it specifies an X axis finish allowance (diameter value)
- **W***n***.** On the second line, it specifies a Z axis finish allowance
- **F***n***.** Feed rate

G73

This example covers the G73 canned cycle being used to machine a profile on a workpiece previously roughed out. See **Figure 15-8**. Note that E designates the position to which the tool must move to begin the first cut. E is dependent on the amount of material to be removed and number of passes to be taken. The operator's manual of a controller should be consulted for a clearer and more detailed explanation of how this position is calculated. The end position of the cycle, when all passes have been taken, is designated by A.

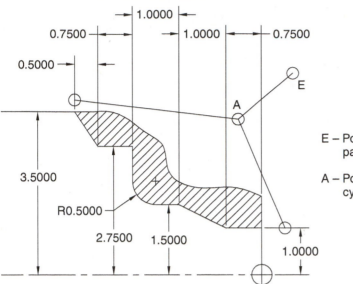

Figure 15-8. Machining a profile on a previously roughed-out workpiece using the G73 canned cycle.

E – Position where first pattern cut begins

A – Point of return when cycle is ended

N70G00X3.350Z-.150 (Position A)
N80G73U.5W.5R.20
N90G73P100Q180U.04W.01F.012
N100G00X2.0Z.100
N110G01Z-.75
N120X3.0Z-1.75
N130Z-2.25
N140G02X4.0Z-2.75R.5
N150G01X5.5
N160Z-3.5
N170X7.0Z-4.0
N180X7.2
N190G00X8.0Z3.0T0500M9

G74—Peck Drilling and Face Grooving Cycle

The G74 cycle can be used for *peck drilling* while making a hole to a certain depth at the center of the workpiece. Remember, the drill does not retract fully after each peck. The drill backs off a distance of approximately 0.005″. Therefore, it will not clear (remove) chips. See **Figure 15-9**. The format for using G74 for peck drilling is as follows:
G74Z*n*K*n*F*n*

Where,

- **G74.** Initiates the cutting cycle
- **Z*n*.** Depth of hole
- **K*n*.** Peck amount
- **F*n*.** Feed rate

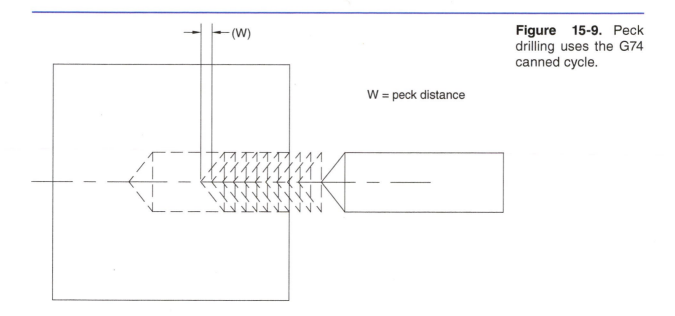

Figure 15-9. Peck drilling uses the G74 canned cycle.

W = peck distance

G74—Peck Drilling

This example covers the G74 canned cycle. It shows the drilling of a workpiece on center with a 0.625″ diameter drill to a depth of 1.250″ and a 0.125″ peck.

```
N60G97S400M3
N70G00X0Z.050
N80G74Z-1.125K.125F.008
N90G00X8.Z5.T0300M9
```

The G74 cycle also can be used to cut two grooves on the face of a workpiece. The format for face grooving is as follows:
G74X*n*Z*n*I*n*K*n*F*n*

Where,

- **G74.** Initiates the cutting cycle
- **X*n*.** Position on X axis
- **Z*n*.** Depth of groove
- **I*n*.** Stepover distance
- **K*nn*.** Peck depth
- **F*n*.** Feed rate

G74—Face Grooving

This example covers the G74 canned cycle. See **Figure 15-10**. It shows a 1.25″ wide groove being cut in the end of a workpiece with a 0.50″ wide grooving tool cutting a groove to a depth of 0.400″.

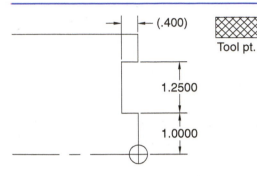

Figure 15-10. The G74 canned cycle can also be used for face grooving.

N130G00X2.0Z.100
N140G74X1.75Z-.400I.350K.050F.020
N150G00X8.0Z5.T0600M9

G75—Cutoff and Grooving Cycle

The G75 cycle is used for cutoff and grooving on both the inside and outside diameters. Peck moves are made when cutting off or grooving. The format for cutoff is as follows:
G75XnInFn

Where,

- **G75.** Initiates the cutting cycle

- **Xn.** X axis move

- **In.** Peck distance

- **Fn.** Feed rate

G75—Cutoff

This example covers the G75 canned cycle. It shows the end of a workpiece being cut off. See **Figure 15-11**.

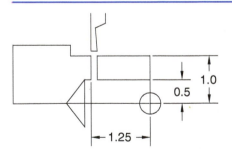

Figure 15-11. Cutting off the end of a workpiece, using the G75 canned cycle.

N190G00X2.1Z-1.25
N200G75X.950I.050F.006
N210G00X8.Z5.T0500M9

Grooving

This example covers the G75 canned cycle. **Figure 15-12** shows grooving on a workpiece. Four external grooves are machined to a depth of 0.5". Groove width is 0.375".

N90G00X4.6Z-1.375
N100G75X3.5Z-5.25I.125K1.25F.005
N110G00X8.Z5.T0300M9
N120.....................

G76—Threading Cycle

The G76 threading command is a two-line command used to machine an entire thread. The command is used for both outside and inside threads. In addition to straight threads, the G76 command is used for machining taper threads. The format for threading is as follows:
G76P*n*Q*n*R*n*

G76X*n*Z*n*R*n*P*n*Q*n*F*n*

G76—Threading

This example covers the G76 canned cycle. It shows a threading cycle that starts from a tool change line. See **Figure 15-13**.

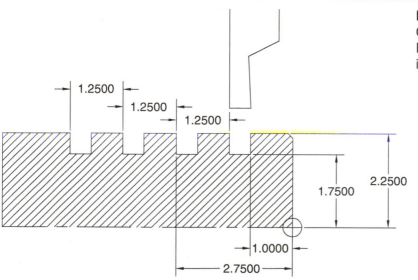

Figure 15-12. Another use of the G75 canned cycle is grooving. Four grooves are being machined in this workpiece.

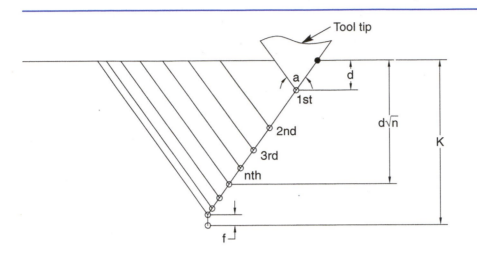

Figure 15-13. The G76 canned cycle is used for threading operations.

N100T0900M8

N110G00X6.0Z6.0T0909M3

N120G00X.7Z.2

N130G76P020060Q10R.001

N140G76X.446Z–.95R0P540Q80F.083

N150G00X6.0Z6.0T0900M9

N160G28U0W0

N170M30

Line N130G76P020060Q10R.001 is read as follows:

- **P020060.** The 02 is finishing passes, 00 is chamfer on thread (not needed), 60 is the angle of tool (60° V thread).

- **Q10.** The 10 is the minimum depth of cut in increments of 0.001", written with no decimals.

- **R.001.** The 0.001 is the finishing allowance or 0.001".
Line N140G76X.446Z–.95R0P540Q80F.083 is read as follows:

- **X.446.** The thread minor diameter is 0.446".

- **Z–.95.** The Z coordinate value of the end of the thread is –0.95.

- **R0.** Indicates a straight thread. If thread is tapered, a value is entered. When cutting a taper pipe thread, the R value is the taper height on one side. In the second G76 line, R should be a minus value R–.05 (R–500). Minus starts low and tapers up.

- **P540.** The total height of the thread is 0.540" (note that there are no decimals used with P-words).

- **Q80.** The depth of the first cut is 0.008".

- **F.083.** The pitch of the thread is 0.083".

Number of threading passes

The formula used to figure the number of passes (see **Figure 15-14)** is as follows:

$$K = \text{total thread depth}$$

$$d = \text{first depth of cut}$$

$$f = \text{amount left for finishing}$$

$$dn = n\text{th depth of cut (if } n \text{ is the number of passes)}$$

$$K = dn + f \text{ (f is assumed to be zero)}$$

$$\text{therefore } n = (K/d)^2$$

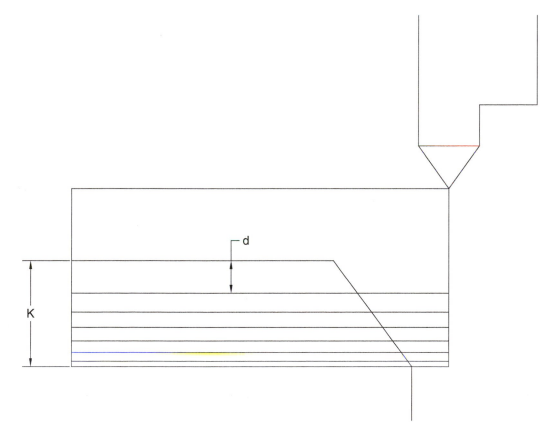

Figure 15-14. Determining the number of passes in a threading operation. The depth of cut is designated by d, while K is the total thread depth.

Example:

Total depth of thread, K = 0.040

first depth of cut = d = 0.018

second depth of cut = d2 = 0.025

third depth of cut = d3 = 0.031

fourth depth of cut = d4 = 0.036

fifth depth of cut = d5 = 0.040

G71, G70, and G76

This example covers a program containing a G71 rough turning command, a G70 finishing cycle command, and a G76 threading command. See **Figure 15-15**.

O2000
N10G20G99G40
N20G96S600M4
N30G50S4000
N40T0100M8
N50G00X6.Z6.T0101
N60X3.2
N70Z.06
N80G01X-.031F.015
N90Z.2
N100G00X3.4
N110Z.025
N120G01X.031
N130Z.2
N140X3.4
N150G71U.0625R.0625
N160G71P110Q280U.03W.03
N170G00X.25
N180G01Z0F.003
N190X.498Z-.125
N200Z-1.0
N210X.630
N220X.750Z-1.12
N230Z-2.628

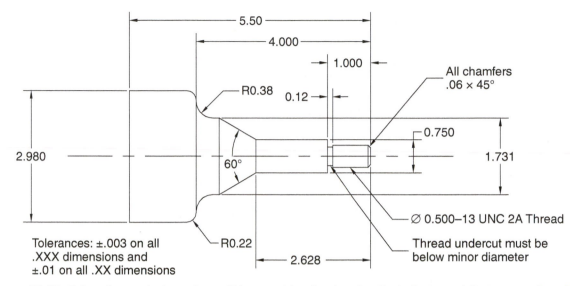

Figure 15-15. Print of a workpiece that will be machined using the G71, G70, and G76 canned cycles.

N240X1.731Z-3.128

N250Z-3.625

N260G02X2.481Z-4.0R.437F.002

N270G1X2.563

N280G03X3.0Z-4.437R.437F.002

N290G00X6.0Z6.0T0100

N300M5

N310M1

N320G28U0W0   *Tool change at home*

N330G96S400M4

N340G50S3000

N350T0700M8

N360G00X6.Z6.T0707

N370X3.2Z0

N380G01X-.031F.003

N390Z.2

N400X3.0

N410G70P110Q280

N420G00X6.Z6.T0700M9

N430M5

N440M1

N450G28U0W0   *Tool change at home*

N460G96S450M4

N470G50S3000

N480T0500M8

N490G00X6.Z6.T0505

N500X1.15Z-1.0

N510G1X.440F.003

N520X1.15

N530G00X6.Z6.T0500M9

N540M5

N550M1

N560G28U0W0   *Tool change at home*

N570G97S800

N580T0900M8

N590G00X6.0Z6.0T0909

N600G00X.7Z.2

N610G76P020060Q10R.001

N620G76X.4041Z-.95R0P470Q.005F.076

N630G00X6.Z6.T0900M9

N640G28U0W0

N650M30

G90, G92, and G94 Canned Cycles

These cycles are useful for roughing and threading. Only the values to be changed need to be specified for repetition of the cycle. Little explanation is necessary to define these cycles. Simple illustrations will be used to show how the cycles work. See **Figure 15-16**.

G90—OD turning

Partial program:

N120G00X4.0Z.250 (Position 1)

N130G90X3.5Z3.5

N140X3.0

N150X2.5

N160G00X6.Z8.

Explanation of a tool path that repeats itself, **Figure 15-17**.
Tool rapids from 1 to 2
Feeds along Z to Position 3
Feed up (X) from 3 to Position 4
Rapids back to Position 1

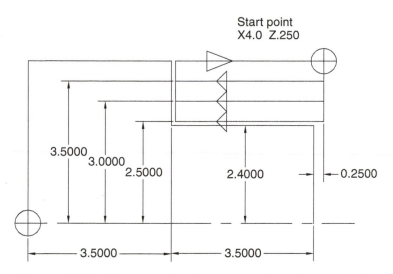

Start point
X4.0 Z.250

3.5000
3.0000
2.5000
2.4000
0.2500
3.5000
3.5000

Figure 15-16. Diagram showing repeated cycles used for roughing and threading a workpiece.

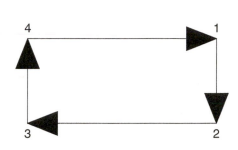

4   1
3   2

Figure 15-17. A tool path that repeats itself in four steps.

G94—Facing

Figure 15-18 shows the operation described in this partial program:

N90G01X3.1Z.1F.015 (Starting point of cycle)

N100G94X1.3Z-.160F.020

N110Z-.320

N120Z-.480

N130Z-.640

N140Z-.800

N150G00X6.0Z3.0

G92—Fixed One Pass Threading Cycle

Using G92 requires the programmer to calculate the depth (X position) of each pass. See **Figure 15-19**. This cycle is seldom used for entire threads because it would be difficult to program the infeed angle (tool cuts on leading angle) move for each thread pass. The G76 command is much better for threading. G92 is usually used as a *spring pass* (a final threading pass to remove any material remaining at the bottom of the thread due to the flex or spring of the material or tool).

- **G92.** Initializes the threading cycle
- **X***n***.** Position on X axis (controls depth of cut)
- **Z***n***.** Length of thread along Z axis
- **F***n***.** Pitch of thread

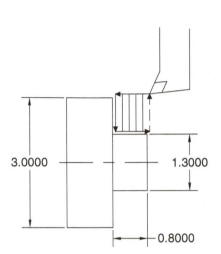

Figure 15-18. A facing operation being carried out with the G94 canned cycle.

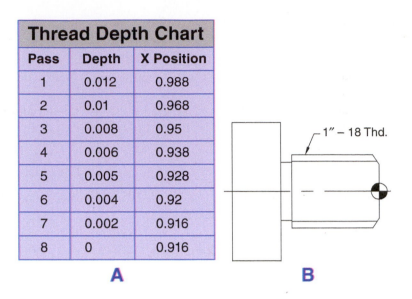

Thread Depth Chart		
Pass	Depth	X Position
1	0.012	0.988
2	0.01	0.968
3	0.008	0.95
4	0.006	0.938
5	0.005	0.928
6	0.004	0.92
7	0.002	0.916
8	0	0.916

A

B

Figure 15-19. The G92 threading cycle. A—Thread depth chart for each pass. B—The workpiece being threaded.

Partial program:

```
N230G00X1.2Z.2
N240G92X.988Z.2F.055          first pass
N250X.968                     second pass
N260X.950                     third pass
N270X.938                     fourth pass
N280X.928                     fifth pass
N290X.920                     sixth pass
N300X.916                     seventh pass
N310X.916                     eighth pass
N320G00X3.0Z3.0
N330M30
```

Summary

The G71 cycle performs turning and boring repetitive commands. The G70 cycle is used to perform the finishing operation after a G71 or G72 rough cycle is performed. The G72 cycle is used to rough face a workpiece. The G73 is a pattern repeating cycle. The G74 cycle can be used to peck drill or perform face grooving. The G75 cycle performs a cutoff operation or grooving on both internal and external diameters. The G76 cycle is an automatic threading cycle.

Simple cycles such as G90 and G94 will perform simple turning, boring, and facing operations. They are called one-pass cycles but can be used to make several passes along a workpiece simply by stating new X or Z values. G92 is a one-pass threading cycle.

Chapter Review

Answer the following questions. Write your answers on a separate sheet of paper.

1. What two letters designate the beginning and end line numbers that define the tool path in a canned cycle?
2. What letter is used to establish the escape or retract amount in a canned cycle?
3. What two letters specify a finish allowance in a pattern-repeating cycle?
4. What letter specifies the peck distance in a G74 cycle?
5. What does the letter *I* represent in a grooving cycle?
6. Explain the meaning of the values in the word P010060.
7. How would you specify a 0.005″ finish allowance in a G76 cycle?
8. What does the letter *K* mean?
9. Why is a G76 threading cycle better than a G92 threading cycle?
10. How much does the drill back off in a G74 peck drilling cycle?

Activities

1. Write the G71 command for rough turning, given the following information:
   * depth of cut 0.200″
   * retract amount 0.030″
   * beginning cycle line number 110
   * end of cycle line number 190
   * finish allowance on diameter 0.020″
   * finish allowance on Z 0.008″
   * feed rate 0.015″
2. Write the first line of a pattern repeating cycle that specifies 0.750″ stock amount on both axes with a retract amount of 0.030″.
3. Refer to the following sketch, and write the G74 command for the partial program below that will machine the face groove, given the following information:
   * 0.375″ wide groove tool
   * pecking distance 0.050″
   * feed rate 0.020″

N150 G00 X4.0 Z.100

N160 G74 _____

N170 G00 X6.0 Z6.0 T0500 M9

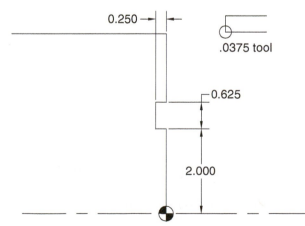

4. Use a G90 cycle to machine the diameter of the workpiece shown below. Use a 0.200″ depth of cut. Write the command. Refer to the example in this chapter if needed.

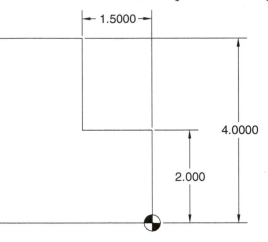

5. Write a rough facing partial program from the sketch below. Use 0.200″ depth of cut. Allow 0.010″ stock for finishing with a feed of 0.007″.

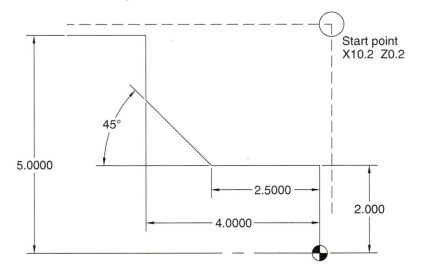

N10G20G99G40
N20G96S800M3
N30G50S4000
N40T0100M8
N50G00X3.35Z1.25T0101
N60G01X3.25F.002
N70G04X0.5
N80X3.35F.05
N90G00X5.0Z0T0101
O1111
N10G20G99G40
N20G96S800M3
N30G50S4000
N40T0100M8
N50G00X3.35Z1.25T0101
N60G01X3.25F.002
N70G04X0.5
N80X3.35F.05

Chapter 16
Subprograms

Objectives

Information in this chapter will enable you to:

- Describe the advantages of using subprograms.
- Write a program using subprograms.

Technical Terms

nesting
subprograms

Subprograms

Subprograms, sometimes called subroutines, are programs called by the main program. A machining operation that has to be repeated on the same job can be used as a subprogram. Rather than rewriting the code for the operation each time it is performed, a subprogram is written and can be called within a main program as many times as needed. Sometimes, there is the need to have subprograms within a subprogram. This is called *nesting*.

Subprograms may contain all the same codes as a regular program, except they must end with a special code (M99). The following is a list of examples where subprograms are beneficial:

- **Making multiple holes with a machining center.** For example, if a workpiece has 25 holes that have to be centerdrilled, drilled, and tapped, commands would have to be given three times to perform those operations. Hole locations would have to be repeated three times because three tools would be used on each hole. With subprogramming, the hole locations have to be written only once within the subprogram, thus saving a lot of programming time and machine memory.

- **Rough and finish contour cuts.** If a complicated-shaped workpiece requires two cuts, a subprogram should be written. This subprogram contains the path coordinates that could be used for both the rough cut and finish cut operations.

- **Machining multiple grooves.** The directions to machine a groove can be put in a subprogram and that subprogram repeated as many times as needed for the number of grooves required.

Subprogram Words

There are four words that are used to command the use of subprograms. These words are M98, P*nnnn*, L2, and M99. M98 commands the control to switch to a subprogram, located in the main program. P*nnnn* instructs the control that O*nnnn* is the subprogram that should be selected. For example, P3000 indicates that the O3000 subprogram will be used. L2 instructs the control to run the subprogram two times. M99 commands the control to return to the main program.

The main program runs until it comes to the M98 command, such as M98P3000. This tells the control to switch to the P3000 subprogram, which takes over and runs until it reaches an M99 command. The M99 tells it to return to the main program and the block immediately after the subprogram block. The main program continues running until it is finished with an M30 command.

Subprogram—Example 1

This example covers a machining center program that centerdrills and drills holes in a workpiece. The coordinates of the holes are found in the subprogram.

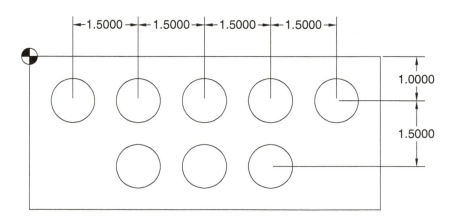

Main Program

O6789

N10G17G20G80G40G49

N20G90G54S5000M3T1M6

N30G00G43H1Z.2

N40G99G81X1.0Y–1.0Z–.25R.2F5.0M08

N50M98P3000L1   *Switch to the O3000 subprogram*

N60G00Z.2M9

N70G91G28Z0G49

N80G17G20G80G40G49

N90G90G54S500M3T2M6

N100G00G43H2Z.2

N110G99G81X1.0Y1.0Z–.75R.2F8.0M08

N120M98P3000L1   *Switch to the O3000 subprogram*

N130G91G28Z0G49

N140G91G28X0Y0

N150M30

Subprogram

O3000

N10X2.5

N20X4.0

N30X5.5

N40X7.0

N50X5.5Y-2.5

N60X4.0

N70X2.5

N80G80

N90M99   *Return to the main program*

Subprogram—Example 2

This example covers subprograms related to turning centers. It shows the machining of several V-grooves in a workpiece. The subprogram makes the moves that actually cut the grooves.

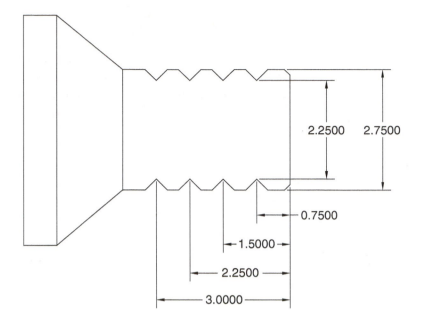

Main Program

O4567
N10G20G99G40
N20G95S400M3
N30G50S3000
N40T0500M8
N50G00X2.9Z-.75T0505   *First Groove*
N60M98P4000   *Switch to the P4000 subprogram*
N70G00Z-1.5   *Second Groove*
N80M98P4000   *Switch to the P4000 subprogram*
N90G00Z-2.25   *Third Groove*
N100M98P4000   *Switch to the P4000 subprogram*
N110G00Z-3.00   *Fourth Groove*
N120M98P4000   *Switch to the P4000 subprogram*
N130G00X3.Z2.T0500M9
N140G28U0W0
N150M30

Subprogram

O4000
N10G01U-.7F.004
N20G04P500
N30G00U.7
N40M99   *Return to the main program*

Summary

Subprograms shorten a program. Any time the program must execute a series of commands more than once is usually a good time to employ subprogramming routines. A subprogam is a separate program containing information that is repeated within a main program.

Chapter Review

Answer the following questions. Write your answers on a separate sheet of paper.

1. What does the letter *L* mean in subprogramming?
2. What is the command to return to the main program?
3. What is a subprogram within a subprogram called?
4. What is the command to go to subprogram O8000?

Activity

1. Rewrite the program for Example 1 in Chapter 10 using a subprogram to shorten the length of program O5 block.

The machine control unit is where the operator sets the workpiece height, tool length offset, and toolholder assignments. (Tibor Machine Products)

N10G20G99G40
N20G96S800M3
N30G50S4000
N40T0100M8
N50G00X3.35Z1.25T0101
N60G01X3.25F.002
N70G04X0.5
N80X3.35F.05
N90G00X5.0Z0T0101
01111
N10G20G99G40
N20G96S800M3
N30G50S4000
N40T0100M8
N50G00X3.35Z1.25T0101
N60G01X3.25F.002
N70G04X0.5
N80X3.35F.05

Chapter 17
Electrical Discharge Machining

Objectives

Information in this chapter will enable you to:

- Describe the operating principle of EDM.
- List several advantages of using EDM.
- List the various components in an EDM system.

Technical Terms

dielectric fluid	electrical discharge	overcut
dielectric solution	machining (EDM)	skim cut
dielectric strength	off-time	slurry
duty cycle	on-time	spark frequency

Electrical Discharge Machining

Electrical discharge machining (EDM) is a process by which metal is removed using electrical energy. An electrode discharges a spark that vaporizes the metal. The electrode and workpiece are both submerged in a *dielectric fluid*, a nonconductive fluid. The basic idea is to move an electrode very close to the workpiece, maintaining a proper distance, and repeatedly produce a spark between both the workpiece and the electrode. See **Figure 17-1.**

A direct current of low voltage and high amperage is applied to the electrode in pulses to create sparks that travel through the dielectric fluid and bridge the gap between the workpiece and the electrode. The *dielectric strength* is the voltage needed to ionize (break down) the dielectric fluid between the electrode and the workpiece.

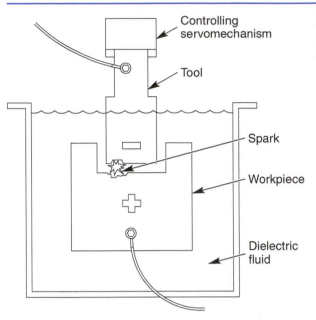

Figure 17-1. This diagram illustrates the basic components of the electrical discharge machining process.

History of Electrical Discharge Machining

Between 1920 and 1940, spark erosion equipment was used as a cost effective way of removing hardened tools, such as broken taps and drills, from expensive machined parts. The early electrodes eroded as much as the metal removed, and more arcing occurred than sparking. Some improvement in efficiency was made when the up-and-down action of the vibrator was used to control spark gap distance. The vibration of the electrode increased the effectiveness of the material removal process.

Two scientists, Dr. B. R. Lazarenko and Dr. N. I. Lazarenko, discovered two improvements that led to a greater usage of EDM in machine shops. The first improvement was the implementation of the RC relaxation circuit, which provided a consistent and dependable control of pulse times. See **Figure 17-2**. The second improvement in the process was the addition of a servo control circuit to automatically find and maintain a given gap. These two improvements led to the EDM machine being used as a production machine tool.

Power supply improvements occurred and led to the development of the transistor. Solid-state components delivered high currents and could switch on-off faster than older vacuum tubes. Controlling the duration of each spark and pause, current amount, and pulse delivery to the electrode has led to the accurate and dependable use of EDM in the manufacturing industry.

Power Supply System

All EDM machines use a pulse-type generator, usually a direct current-type that activates and deactivates the current in high frequency pulses. The number of pulses per second is referred to as the *spark frequency*. The

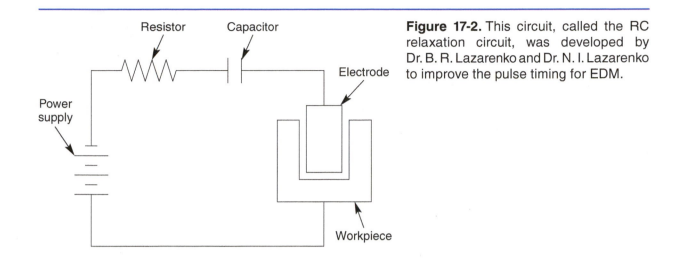

Figure 17-2. This circuit, called the RC relaxation circuit, was developed by Dr. B. R. Lazarenko and Dr. N. I. Lazarenko to improve the pulse timing for EDM.

power supply is microprocessor controlled and solid-state. See **Figure 17-3**. The period of time in which current is activated is called *on-time*, and the time between the current pulses is called *off-time*. The percentage relationship between the on and off times of current flow is called the *duty cycle*. For example, a cycle of 100 millionths of a second and a current on for 45 millionths of a second and off for 55 millionths of a second is considered to be a 45 percent duty cycle.

Flushing System

In sinking EDM machines, a paraffin, kerosene, or silicon-based dielectric fluid is used to prevent premature spark, cool the workpiece and electrode, and flush or remove particle debris. The dielectric tank is

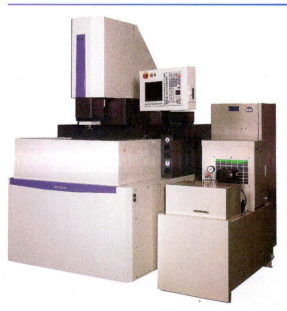

Figure 17-3. The power supply on today's EDM is controlled by a microprocessor, ensuring accurate pulses of current.

attached to a dielectric pump, an oil reservoir, and a filter system. The pump furnishes pressure for flushing the work area and supplying the oil, while the filter system removes and captures debris in the oil. The oil reservoir stores additional oil. In wire EDM machines, deionized water acts as a dielectric fluid. Dielectric fluids are cooled in heat exchangers and filtered to remove dislodged debris that could reduce efficient flushing in the spark gaps.

Flushing is accomplished by directing the fluid to flow through the spark gap under pressure, by sucking it through the gap, or by allowing a nozzle to agitate the tank fluid that surrounds the workpiece. See **Figure 17-4**. Pressure flushing channels fluid through holes placed in the electrode or the workpiece.

It is extremely important to remove debris completely during the cutting process for successful EDM machining. Entrapment of gases caused by sparking should be prevented since the explosion of these gases could lead to personal injuries, electrode damage, or fire.

Care should also be taken when fluid travels through narrow passages. High fluid pressure might lead to electrode and workpiece movement, causing inaccuracy to the finished workpiece.

Electrodes

Conventional (ram) type EDM machines use electrodes made from graphite. This material is sold in blocks and cylinders, and is machined to a positive shape. See **Figure 17-5**. This shape produces a negative shape (cavity)

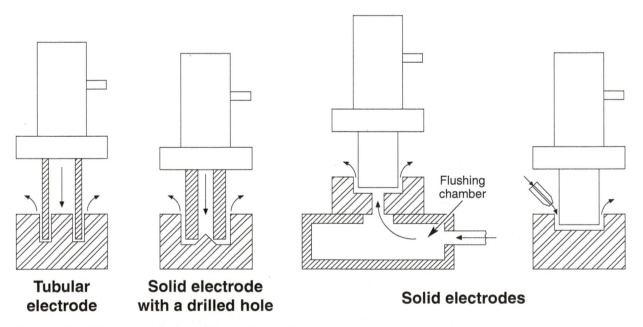

| **Tubular electrode** | **Solid electrode with a drilled hole** | **Solid electrodes** |

Flushing chamber

Figure 17-4. The metal fragments are flushed out of the cutting area by pumping fluid through the electrode, through the workpiece, or into the cutting area externally.

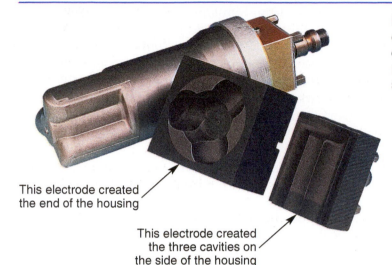

Figure 17-5. Blocks of graphite, or other electrode materials, are shaped using conventional machining processes and mounted on an EDM to produce the final shape in metal. (GF AgieCharmilles)

This electrode created the end of the housing

This electrode created the three cavities on the side of the housing

in the workpiece. At low frequencies, graphite has a high resistance to heat and wear. However, it wears more rapidly at a high frequency or with the use of negative polarity. Because electrodes wear fairly rapidly, a number of electrodes may be necessary to complete one job. Other electrode materials include copper-graphite alloys, copper, tellurium copper, brass, tungsten, and several tungsten alloys. Electrodes are shaped using conventional machining practices. Electrode material is selected on the basis of metal removal rate, wear resistance, desired surface finish, electrode cost, and characteristics of the metal being cut.

Electrode wear

Electrode wear can be reduced by maintaining a small amount of finish material on the workpiece. Enough material should be left merely to remove any craters caused by the roughing process. Use low power with high frequency and minimum offset for finishing passes.

Overcut

Overcut is the distance between the side of the electrode and an adjacent cavity wall. This distance increases with longer on-time, greater spark energy, and higher amperage. Electrode size allowance must be considered to control overcut. For example, electrodes must always be manufactured smaller to avoid an oversized cavity or hole.

Surface Finish

High spark frequency is used for finishing operations and for cutting cemented carbide, titanium, and copper alloys. High spark frequency with small gaps results in finer finishes. Low spark frequency with large spark gaps removes metal rapidly and results in a rough finish. Electrode wear increases with high frequency and decreases with low frequency.

Normal oil is used for roughing and finishing cuts. A *slurry* consisting of a silicon powder added to the dielectric fluid (oil) can produce a super matte and highly reflective finish. A *dielectric solution* is a nonconductive liquid that fills the gap between the electrode and workpiece and acts as an insulator until a specific gap and voltage are achieved. The solution then ionizes and becomes an electrical conductor, allowing a spark (current) to flow through it to the workpiece causing removal of metal. The dielectric solution also cools the workpiece and flushes away the particles generated by the spark.

Dual tanks on machines allow standard machining and silicon powder operation. Good detail can be produced ten times faster than with general machining. Normal sparks are reduced and multiplied through the use of silicon powder, which act as semiconductor particles. Both the standard machining side and the silicon powder slurry side have separate chillers to maintain proper thermal stability of the entire machining process. Less than .0003" of the material is removed with powder. With the silicon powder operation, polishing is eliminated, spark penetration is greatly reduced, less energy is used to produce final surface, and no flush is necessary. This leads to a more uniform surface finish. However, keep in mind that a powder slurry should only be used when a superior finish is needed. Use of the powder restricts maximum power and reduces powder life. Powder oil cannot be filtered.

Surface roughness is the measure of surface irregularities. Surface roughness (Ra) is rated as the arithmetic average deviation of surface valleys and peaks, expressed in microinches. A 16 Ra finish is equal to a surface finish produced by burnishing, grinding, honing, or polishing. Features created by EDM have an "orange peel" appearance. A finish of 16 Ra is achievable with EDM, but 64 Ra or higher is more common and less expensive to produce.

Advantages of EDM

In the past, EDM was primarily used to produce parts that were difficult to manufacture using conventional processes. EDM is no longer the last choice for machining: instead it is often the first choice for designers. Some of the advantages EDM has over other machining processes include the following:

- It is possible to machine hardened materials
- Broken taps and fasteners are easily removed
- No part distortion on thin workpieces
- Intricate shapes can be machined
- Outstanding finishes can be produced
- One operator can run several machines
- Ram machines cut carbide quicker

Disadvantages of EDM

- EDM is not a fast method of machining
- Nonconductive materials such as plastics, glass, or ceramics cannot be machined

Wire EDM

Wire EDM uses a continuous feeding wire electrode that is guided and monitored by a computer numerically controlled (CNC) system. See **Figure 17-6**. The wire is fed from reel to reel and travels through the workpiece. Rapid dc electrical pulses are generated between the wire electrode and the workpiece. A shield of deionized water, called the dielectric, is passed into the gap to help ensure good spark integrity. The water also helps cool the gap and flush away debris after each spark. Deionized water (distilled water), which is a very pure water, lowers the conductivity of water, making it suitable for flushing debris during metal erosion when cutting with a wire electrode. The most important factor in maintaining accuracy and stability is keeping a consistent spark gap between the wire and workpiece.

The process for wire EDM is as follows:

1. A conductive workpiece is fastened to the machine table. A certain amount of voltage is applied to the wire as it is moved within close proximity of the workpiece. A dielectric fluid (deionized water) is forced between the two surfaces.

2. As the wire is moved toward the workpiece, the open voltage creates an electromagnetic field between the two surfaces. The magnetic field pulls the ions and conductive debris from within the fluid to form bridges across the gap.

3. As these bridges become denser, the resistance decreases and eventually the resistance of one of these bridges will drop below the threshold and an extremely hot spark (current) will jump across that ionic bridge. The temperature of the spark is estimated to reach 20,000°F.

4. At the initiation of the spark, a certain amount of material is vaporized on both surfaces. This vaporization accounts for approximately 30% of the total erosion process. While the spark continues to short across the gap, the intense heat starts to break down and boil the material in the surrounding area. Also, because of the heat, there is a rapid expanding pressure around the entire spark. The spark is small, but with the heat, surrounding agitation, and pressure, it is intense. This period of time in which the current is arcing across the gap is known as on-time. Because of polarity (direction of current flow), the EDM process always takes a larger "bite" out of the workpiece than the wire. This helps to ensure that the wire will not be burned in half.

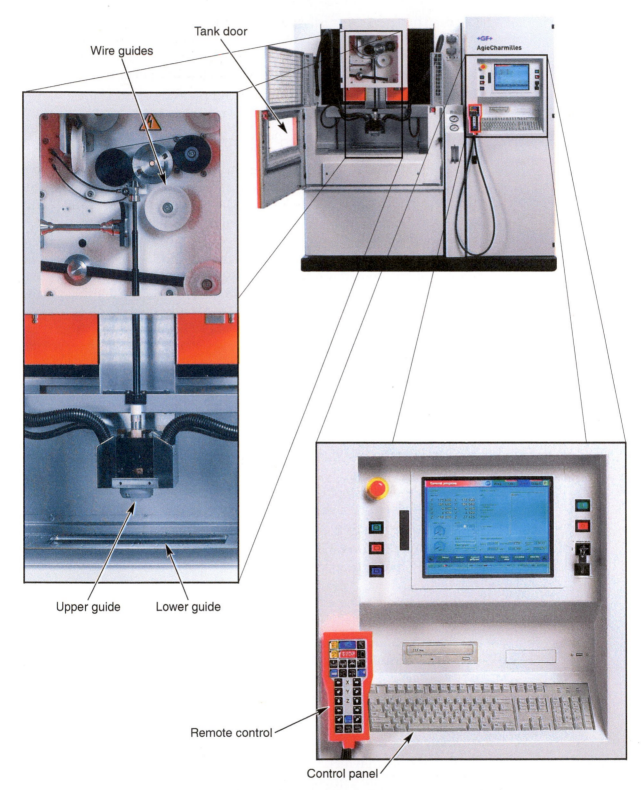

Figure 17-6. This model is typical of wire EDM machines. (GF AgieCharmilles)

The on-time proceeds for a specified period of time (controlled by the electrical settings), after which the current is shut off. This is the initiation of "off-time," a specified period of time in which the current is completely removed. All EDM processes use this current pulse action.

5. At the beginning of off-time, the heat and pressure sources are removed. This causes a strong implosion or vacuum in the spark area.

6. This implosion forces most of the boiling molten material to be sucked away from both the electrode and the workpiece into the vacuum. This molten metal is cooled and solidified into small hollow spheres. These particles and the ionic bridge are then flushed away prior to the next spark. This accounts for the other 70% of the erosion process. Water itself is not conductive, but the high amount of ions (conductive particles) that water usually contains are. If water is completely stripped of ions (distilled water), it becomes a perfect insulator and no spark could jump the gap. If there are too many ions making the water highly conductive, the current will prematurely draw across the gap, causing the loss of many potential sparks to a direct short condition. The result in the burn would be fewer and less powerful sparks reducing efficiency. To avoid this, the amount of ions is decreased to a small amount, using deionizing resin. This helps resist a spark until the voltage reaches full potential.

Types of Wire EDM

There are three types of wire EDM machines: a two-axis machine, a simultaneous four-axis machine, and an independent four-axis machine. The two-axis machine makes right angle cuts. The simultaneous four-axis machine is able to produce tapers, as well as a combination of taper and straight surfaces. The independent four-axis machine has the capability of having the top profile of the workpiece different from the lower profile of the workpiece. See **Figure 17-7**.

Finishes, Accuracy, and Tolerances

Finishes below 16 Ra are easily obtainable on wire machines. Machines can hold a tolerance of +/–.0002″ on thick workpieces and tolerances of +/–.0001″ on thin pieces.

Skim cuts

To increase surface finish quality and keep dimensions within tight tolerances, the part's edge may be run close to the wire to remove a small amount of material; this is called a *skim cut*. Skim cuts are needed where cornering has produced an overcut or undercut on sharp corners, because of dwells on inside radii and speeding on outside radii. To increase accuracy, a skim cut is taken to obtain straight sides. Also, stresses in the metal may cause the part to move therefore a skim pass is taken to increase part accuracy.

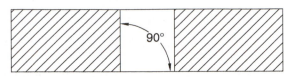

Right angle cut with a two-axis machine

Taper cut with a simultaneous four-axis machine

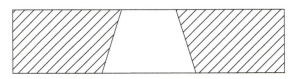

Taper and land cut with a simultaneous four-axis machine

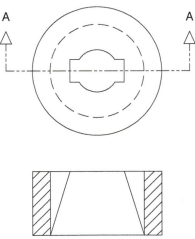

View A-A

Multiple profiles cut with an independent four-axis machine

Figure 17-7. The three types of wire EDM machines—two-axis, four-axis, and independent four-axis—are able to cut increasingly complicated profiles.

Flushing

Flushing is important when using wire machines. Top and bottom flushing nozzles provide flushing pressure that helps produce the spark and remove dislodged metal particles. The gap between the workpiece and nozzle is important to the process. Difficulty in cutting can arise when the nozzle gets close to the edge of the work and flushing pressure is decreased. An attempt should be made to keep the flushing nozzles a minimum of 3/16" from the edge of the material.

Wire Speed

Wire speed will vary depending upon the power level, material thickness, and surface finish results desired. Speed varies from notch 1 to notch 16 (9 linear inches per second). Notch numbers refer to the settings of the wire speed rates. As the wire erodes while cutting, it loses size and strength. However, if the wire speed is too low, the wire may break and cause bouncing contact (lines in work) or straightness deviation. These problems would negate any savings in running wire speed lower. Wire speed should be increased on skim cuts because of the very high frequencies that cause high wire wear.

Wire Tension

Wire tension is set according to the wire type and diameter, as well as the type of cut. Skim cuts require higher tension than rough cuts. High-speed cuts require the wire tension to be reduced to avoid wire breakage. Generally on rough cuts, notch tension runs from 1 notch to 16 notches (100 grams to 2500 grams). Wire tension notch is set equal to the wire size for most cuts. For example, if the wire diameter is .008", then the wire tension notch 8 is used. Skims cuts are 3 notches or 4 notches higher than the wire diameter.

Fluid Pressure

Generally, low pressure is used for skim cuts and high pressure is used for rough cutting. Rough cuts may use 8 to 10 liters per minute, while skim cuts use approximately 1 liter per minute. Water pressure should be lowered if part vibration occurs.

Liquid Resistivity

The higher the notch number, the more conductive the water. Ions make water conductive. High ion content makes water too conductive, causing premature sparking and dense concentrations of sparks in the machining gap. This could lead to wire breakage, rough finishes and inconsistent gap widths.

Summary

EDM removes metal by discharging sparks that vaporize the metal. A dielectric fluid is used to prevent premature spark occurrence, cool the workpiece and electrode, and flush or remove particle debris. Electrodes are commonly made from graphite and machined to a positive shape that conforms to the cavity shape required. Surface finishes produced by EDM are outstanding. EDM can machine hardened materials.

Wire EDM uses a continuous feeding wire electrode that travels through the workpiece, guided and monitored by a CNC system. Wire EDM machines can handle up to five axes of movement.

Chapter Review

Answer the following questions. Write your answers on a separate sheet of paper.

1. State how EDM works.
2. What is a *dielectric fluid*?
3. What is *duty cycle*?
4. What type of dielectric fluids are used in sinking machines?
5. What materials are used to make electrodes?
6. What type of power and frequency are used for finishing passes?
7. List at least three advantages of EDM.
8. State a disadvantage of EDM.
9. List three types of wire EDM machines.
10. What settings must be made to wire on EDM machines?
11. When are skim cuts needed?
12. What tolerances can be held on wire machines?
13. When is low fluid pressure used?
14. What is accomplished by increasing the notch value on a wire?

Activities

1. Prepare a report that explains the advantages and disadvantages of the different electrode materials available for sinking EDM.
2. Select an industry and identify five products, fixtures, or tools that are produced using EDM.
3. Search the Internet for EDM information.

Glossary

A

A axis. A tilting of the spindle on a five-axis machining center. (4)

absolute coordinates. A method of plotting points that use the origin point as the reference point. (2)

absolute measuring system. Uses measurements from a fixed reference point to specify a point or location. (2)

acute angle. Angle that is greater than 0° and less than 90°. (3)

adapter mounting. The cutter is mounted on a piloted adapter that uses drive keys. The cutter is held in place by a lock screw or socket head cap screw. (6)

adjacent angles. Two angles that use a common side. (3)

angle. The figure formed by the meeting of two lines at the same point. (3)

angle plates. Used to hold irregularly shaped work with a flat surface, or hold work that must be held at a right angle to the axis of tool travel. (4)

arc. A segment of a circle. (3)

attribute data. A quality characteristic that can be observed as present or absent, complete or incomplete, within tolerance or out of tolerance, conforming or nonconforming. (18)

audit. A procedure designed to investigate the activities, practices, records, and policies of an organization to determine if the organization has the ability to meet or surpass a standard. (19)

B

B axis. This is the axis created when using a rotary table. (4)

ball screw. A hardened and ground precision lead screw that has a recirculating ball bearing nut used to convert rotating motion to linear motion. (1)

ball-nose end mills. End mills ground with a male radius on the end. (5)

bar pullers. Used to clamp and pull bars of stock through spindle for machining. They are mounted in the turret. (11)

bed. Supports the major components of a machine. It can be flat or slanted. (11)

binary system. Uses only the digits zero (0) and one (1). This system is basically structured the same as the decimal system, but is constructed on the powers of base 2 and not base 10. Also known as *base 2 system* (1)

binary-coded decimal system (BCD). Uses the decimal system for place locations but converts each digit of the decimal number into a binary number. (1)

bisect. To divide into two equal parts. (3)

block. A group of words that tells the computer to act on a complete statement of instructions. (8)

boring. An internal turning operation used to finish a hole to size and remove any imperfections. (11)

boring canned cycle. Cycle used when a drag line (mark) is allowed in the hole caused by the tool rapiding out of the hole when the spindle stops. (10)

brainstorming. A technique of soliciting a list of problems or ideas from a group of people. Everyone should participate and should be given an opportunity to speak. (18)

C

canned cycles. A method of shortening the length of programs while performing such operations as drilling, centerdrilling, reaming, spotfacing, counterboring, tapping, and boring.

carbide. A term commonly used to refer to cemented carbides. (5)

carousel. Storage system found on vertical machining centers. Tools are loaded in a coded drum. The drum rotates with the desired tool to a tool change position. (4)

Cartesian coordinate system. Lines or planes that run horizontally and vertically and intersect with each other. These lines or planes are perpendicular to each other. A third line or plane can be added to this system and is perpendicular to the other two. With Cartesian coordinates, any point in space can be described in mathematical terms from any other point along three mutually perpendicular (90°) axes. (2)

center drill. Used to produce accurate starter holes so that drills will begin in perfect alignment and location. (5)

ceramic tools. Tools made primarily from aluminum oxide, sometimes with titanium, magnesium, or chromium oxides added. Ceramics are quite brittle and require very rigid tooling and setups. (5)

cermets. Ceramic/metal composites comprised mainly of titanium carbide (TiC) and titanium nitride (TiN). (5)

certification. A written statement by an official group that attests that specific equipment, processes, or a person complies with a specification, contract, regulation, or policy.

chip conveyors. Used to remove chips from a machine. (11)

chord. A straight line that joins any two points on the circumference of a circle. (3)

chucking reamers. Straight- or taper-shank one-piece reamers with either straight or spiral flutes. (5)

circle. A set of points, located on a plane, that are equidistant from a common central point (center point). (3)

circumference. The distance around a circle. (3)

clamps or **straps.** These include plain bar clamps, spring-loaded L-clamps, U-clamps, offset clamps, and universal clamps. (4)

climb milling. Feeding the work in the same direction as the rotation of the cutter. It is commonly used on CNC machining centers. Climb milling has cutter forces acting in a downward direction, which tends to pull the work into the cutter. Also called *down milling* (5)

closed-loop system. Positioning control system that uses resolvers to continually monitor spindle and table movement and report this data back to the MCU, comparing the current position with the programmed position. (1)

coarse feeds. Feeds used for rough cutting and generally softer materials. (12)

collet chuck adapter mounting. The collet reduces runout. This provides better part finishes and size control. (6)

collet chuck. CNC turning center component that clamps the workpiece at nearly every point around its circumference. It is usually used with bar feeders and is programmable. (11)

complementary angles. Two angles that equal 90°. (3)

computer numerical control (CNC). The process by which a computer controls the operation of a machine tool. (1)

computer-aided manufacturing (CAM). Software that allows the programmer to create part geometry, select and create tooling, determine machining operations, and create tool paths. (1)

congruent. Having the same size and shape. (3)

contour turning. Various operations such as chamfering, turning, recessing, grooving, tapering, and radius forming. These operations are combined to generate a specific shape to the workpiece. (11)

conventional milling. Feeds the work against the rotation of the cutter. Cutting forces are directed upward. It is most commonly used on manual or conventional milling machines. Also called *up milling* (5)

coolant. The fluid used during machining to reduce heat caused by friction and to remove chips. (4)

core drills. Three- or four-flute drills used to enlarge existing cored, drilled, or punched holes. (5)

corner radius end mills. End mills that produce male radii. (5)

cosecant. The ratio of the hypotenuse to the opposite side. It is the reciprocal of the sine function. (3)

cosine. The ratio of the adjacent side to the hypotenuse. (3)

cotangent. The ratio of the adjacent side to the opposite side. It is the reciprocal of the tangent function. (3)

counterboring. The operation of enlarging the upper end of a hole to produce a flat bottom portion so the head of a machine screw or bolt is below the surface. (5)

counterboring canned cycle. Almost the same as the spotface, counterbore canned cycle. This cycle is used with a boring bar; and the spindle stops before rapiding out of the hole. It is used to produce a smooth hole to a precise depth. (10)

countersinking. The operation of enlarging the upper end of a hole into a tapering, or conical, shape. (5)

cutter pitch. The number of inserts found on a milling cutter. (6)

cutting speed. A measurement of the distance (in feet or meters) that the circumference of the work passes the cutting tool in one minute. Cutting speed is expressed as surface feet per minute (sfpm) or surface meters per minute (smm). (5)(12)

D

datum. The point at which two axes intersect. Also called *origin* or *zero point* (2)

decimal system. System which uses base 10 or the power of ten to indicate a numerical value. (1)

depth of cut. The distance the tool is fed perpendicular into the workpiece. It determines the amount of material being removed. (12)

diagonal. Running from one corner of a four-sided figure to the opposite corner. (3)

diameter. The length of a straight line that connects two points on a circle and intersects through the center of the circle. (3)

diameter programming. Programming in which a diameter is specified. (13)

dielectric fluid. A nonconductive fluid in which the electrode and workpiece are both submerged. (17)

dielectric solution. A nonconductive liquid that fills the gap between the electrode and workpiece and acts as an insulator until a specific gap and voltage are achieved. (17)

dielectric strength. The voltage needed to ionize (break down) the dielectric fluid between the electrode and the workpiece. (17)

direct numerical control (DNC). A system using a computer that is hard-wired to several machines. The computer controls the operation of all machines in its network. Programs can be loaded from a shop office directly to a machine on the shop floor. (7)

drawbar. A device that grabs and pulls the toolholder retention knob. (4)

drill, ream canned cycle. A simple cycle that feeds the tool to a specified depth then rapids out of the hole. (10)

drilling. Used to produce a hole in the workpiece. (11)

drive motor. Controls the machine slide travel on NC/CNC machines. (1)

duty cycle. The percentage relationship between the on and off times of current flow. (17)

dwell. A command used by the machining center to pause tool movement at a particular time in a part program. (4)

dwell command. Causes axis motion to pause for a specific amount of time. It is frequently applied in situations where tool pressure needs to be relieved. (8)(14)

dwell command. Stops all X and Z movement. It is primarily used to break metal chips at the end of a machining operation, such as necking, grooving, boring, or turning to a shoulder. (8)(14)

E

electrical discharge machining (EDM). A process by which metal is removed using electrical energy. An electrode discharges a spark that vaporizes the metal. (17)

electrical numerical integrator and calculator (ENIAC). The world's first digital electronic computer, created in 1945. (1)

end mill adapter mounting. The cutter is held with a lock screw against a Weldon shank. (6)

engineering drawing. A graphical description of a workpiece. It contains dimensions and tolerances. (7)

equilateral triangle. Triangle with three equal sides. All angles are equal (60°). (3)

external toolholder identification system. ANSI standards used to identify external toolholders. (12)

F

face milling. The process of creating a flat surface using an end mill or face milling cutter. (5)

face mills. Manufactured with HSS, brazed carbide, and indexable carbide inserts. (5)

facing. Removing material from the end of the workpiece. (11)

F-command. Used to set the feed rate on a CNC machine. (4)

feed. The distance that the work advances into the cutter. It is measured in inches per minute (ipm) or millimeters per minute (mpm). (5)(12)

fine feeds. Feeds used for finishing and for cutting harder materials. (12)

finishing cuts. Used to size the workpiece and provide a good surface finish. (12)

first zero. Machine zero position located to the extreme right-hand corner of the machine. (14)

fixture offsets. Work coordinate settings that assign program zero to a workpiece. (8)

fixtures. Specially designed holding devices used exclusively to hold and locate a specific part or parts. (4)

flowchart. Shows the programming sequence used to manufacture a workpiece. (7)

fluteless taps. Used to form a thread rather than cutting a thread. They form threads in the hole with lobes on the outside edge of the tap. (5)

function. A magnitude (size or dimension) that depends on another magnitude. (3)

G

G-codes. Codes that determine the conditions under which the machine will function. These codes direct the equipment through various machining operations in addition to specifying axis moves. Also called *preparatory codes* (8)(14)

gang-style toolholder. Toolholder used on small CNC machines. It has tools mounted on a table. (11)

graphical proveout. A visual representation of a cutting tool, the tool path, and the material that is being cut. (13)

grooving. Grooving is accomplished using a tool that is fed perpendicular to the workpiece centerline to a specified depth. It is often used to provide relief for a threading operation. (11)

H

headstock. CNC turning center component that contains the spindle and spindle drive motor. (11)

high-helix drills. Drills that have a high angle (35°–40°) on the flutes. This angle assists in removing chips more rapidly when machining deep holes. (5)

high-speed steel (HSS). An alloy steel that has the ability to resist wear and to retain strength at high temperatures. (5)

home position. Used as a reference point to locate or set the program zero point. (2)

home. Machine zero return. (8)

horizontal machining center. Machining center that has the machine spindle mounted with its axis in a horizontal position. It is much more flexible than the vertical machining center, but more difficult to operate and set up. (4)

hypotenuse. The side of the triangle that is opposite the right angle. (3)

I

IJK method. Method of programming a circular move. (8)

incremental coordinates. Use the current position as the reference point for the next move. When incremental coordinates are used with turning centers, the addresses U and W are usually used to replace the addresses X and Z. (2)

incremental measuring system. Uses a floating reference point to specify a point or location. Each new tool location and movement uses the last location as a reference point. (2)

indexable cutting inserts. Disposable inserts made of various materials and manufactured in a number of shapes. The objective is to provide an insert with several cutting edges. (2)

indexable insert drills. Drills that range in size from 5/8"–3". They are similar to spade drills but use replaceable indexable carbide inserts. (5)

indexers. Devices that allow the workpiece to be quickly rotated a specified number of angular derees. (4)

initial plane. The last Z position of a tool prior to the canned cycle command. (8)

insert grades. Carbide insert grades are based on their toughness and their ability to resist wear. Different grades are also based on the type of cuts being taken. (6)

insert size. Determined by the largest inscribed circle (IC) that will fit inside the insert or touch all edges of the insert. (6)

inspection sheet. Provides the operator with information on the specific dimensions or features to check, tolerances on these dimensions or features, and the inspection devices or equipment that should be used. (7)

isosceles triangle. Triangle with two equal sides and two equal angles. (3)

L

lead angles. Lead angles on a milling cutter have an effect on cutting force direction, chip thickness, and tool life. Lead angles can be 0°, 15°, 20°, and 45°. (6)

live tooling. Tools that can rotate while mounted in the turret. (11)

low-helix drills. Drills that have a low angle on the flutes. The low angle prevents the drill from digging into soft material too quickly. (5)

M

machinability. The difficulty or ease with which a metal can be cut. (5)

machine body. Framing consisting of a bed, saddle, column, and table. (1)

machine control unit (MCU). Computer that interprets alphanumeric characters that make up the series of sequenced instructions used to operate a machine tool. It has a control panel containing switches, buttons, and a monitor screen. (1)(11)

machine home. Zero position. (14)

machine screw clamping. Method used to fasten the insert to the holder using a fine thread button head screw. (12)

machine spindle. A device that holds cutting tools (on a CNC mill) or workpieces (on a CNC turning center). (1)

machine zero. A very accurate position along each of the machine's axes that is set by the machine tool builder. Upon machine startup, the machine axes must be sent to the zero return position before operating the machine. (2)

machining center. A milling machine that uses programmed commands, has computer controlled movement in three or more axes, and is able to perform automatic tool changes. (4)

manual data input (MDI). Direct method of loading program data into the memory of the CNC machine. It is performed at the machine, using the controller keyboard. (4)(7)

manufacturing instruction sheet (MI). Procedure sheet containing elements such as part name, part number, machine name, material specifications, fixture or holding device identification, and columns containing sequence numbers and descriptions of each sequence. Also called *manufacturing methods sheet* (13)

M-codes. Control miscellaneous functions in a program. (8)

milling. The process of removing material with a multitooth rotating cutter. (6)

modular fixtures. Specialty components designed for clamping. They are assembled on a special plate that has accurately located tapped holes to which components are attached. (4)

multiple clamping. Method used to fasten the insert to the holder using both a clamp and eccentric pin. (12)

N

negative rake inserts. Inserts used for general purpose machining, the majority of steels, and cast irons. (12)

nesting. Subprograms within a subprogram. (16)

nose radius. The larger the nose radius, the stronger the insert and the better the finish. However, if tooling and setup are not rigid, a large nose radius can cause chatter. (6)

numerical control (NC). The operation of a machine tool using a series of sequenced instructions consisting of alphanumeric characters (letters and numbers). (1)

numerical control programming. The process of combining print information, tooling information, setup information, and speeds and feeds with a sequence of operations. (1)

O

obtuse angle. Angle that is greater than 90° and less than 180°. (3)

off-time. The time between the current pulses. (17)

oil hole drills. Twist drills with one or two oil holes running from the shank end of the drill to the cutting point. (5)

on-time. The period of time in which current is activated. (17)

open-loop system. Positioning control system used with stepper motors. It works on the principle of rotary movement of 1.8° for each electrical pulse received. (1)

origin. The point at which two axes intersect. (2)

overcut. The distance between the side of the electrode and an adjacent cavity wall. (17)

P

parallel. Lying in the same direction but always the same distance apart. (3)

parallelogram. A quadrilateral with equal opposite sides and equal opposite angles. (3)

parallels. Precision-ground flat, square, or rectangle steel or cast iron bars, usually used in vises to support and raise a workpiece. (3)(4)

part catchers. Programmable devices used to catch parts as they are cut off the bar stock. (11)

parting. Used to remove the finished workpiece from the bar stock. The tool is fed perpendicular to the bar stock until the workpiece is separated from the stock. (11)

peck drilling. A drilling procedure that retracts the drill slightly after drilling a portion of the total intended depth. (15)

peck drill canned cycle. This cycle is used to remove chips when drilling deep holes. (10)

perpendicular. At a right angle to a line or surface. (3)

pin lock clamping. Method used to fasten the insert to the holder. An eccentric pin holds the insert in place. (12)

plug tap. A type of hand tap that is frequently used in machine tapping on CNC machines. (5)

polar angle. The angle formed by the vector and the polar axis (X). (2)

polar axis. A horizontal reference line drawn from the origin out to the edge of a circle. (2)

polar coordinates. Define the position of a point by its distance and direction from a fixed reference point that has a value of zero. (2)

polygons. Figures with many sides that are formed by line segments. (3)

positive rake inserts. Inserts that help cut down on chatter or part bending. They are used when machining long, slender, or thin-walled workpieces and on soft steels and nonferrous metals. (12)

preparatory codes. Codes that determine the conditions under which the machine will function. These codes direct the equipment through various machining operations in addition to specifying axis moves. Also called *G-codes* (14)

profiling. The process of contouring a workpiece with a cutter that has teeth on its outside edge. (5)

profiling cycle. Cycle that allows the repeated cutting of a fixed pattern, with the pattern being shifted little by little. (15)

program. A hard copy of the G-code program. (7)

program documentation. The paperwork used for a CNC job. This may include the engineering drawing, setup sheet, tool list, tool library, and inspection sheet. (7)

program manuscript. The written CNC program containing the preparatory and miscellaneous codes. It also contains the axis information, feed rates, and spindle speeds. (13)

program zero. A floating point (X0,Y0) that can be set at any position inside the machine's grid system limits. It can be located anyplace on the part to establish a start point from which you may calculate the rest of the coordinates. (2)

proposition. A statement to be proved, explained, or discussed. (3)

Pythagorean theorem. States a special relationship that exists among the three sides of a right triangle. It states that *the length of the hypotenuse squared equals the sum of the squares of the other two side lengths.* (3)

Q

quadrants. The four sections of a two-axis coordinate system. (2)

quadrilateral. A polygon with four sides. (3)

qualified. Toolholders are labeled "qualified" when the dimensions from the tip of the insert to the side of the toolholder and the end of the toolholder are within a tolerance of ± 0.003". (12)

R

radius. A segment that joins the circle center to a point on the circle circumference. (3)

radius method. Method of programming a circular move. (8)

radius programming. Programming in which radius value is specified. (13)

reamer. Multiflute tool used to finish a hole to size. (11)

reaming. A finishing and sizing operation performed on a drilled or bored hole using a multiflute tool. (5)(11)

reaming canned cycle. Cycle in which the tool *feeds* out of the workpiece instead of *rapiding* out. (10)

rectangle. A quadrilateral with equal opposite sides and four right angles. (3)

reflex angle. Angle that is greater than 180° and less than 360°. (3)

registrar. An individual or firm responsible for the review and analysis of a program dealing with quality compliance. (19)

relief angle. This angle may be a combination of the insert and toolholder. It can vary from 0°, which is used with negative rake holders, to 25°. Also called *clearance* (12)

right angle. Angle that is exactly 90°. (3)

right triangle. Triangle that has a 90° (right) angle. (3)

root mean square. Relates to surface roughness average and is used to describe surface finish produced by a metal cutting operation. (5)

rotary tables. Rotating devices that are much more flexible than indexers. The rotation angle can be controlled more precisely. (4)

roughing cuts. Used to remove excess material without regard to surface finish. (12)

roughing end mills. End mills with grooves or scallops around the periphery. (5)

R-plane. The R value found in the canned cycle command. (8)

rpm (revolutions per minute). The number of spindle rotations in one minute. (12)

S

scalene triangle. Triangle with three unequal sides and unequal angles. (3)

scaling. Multiplies the program's X, Y, and Z values by a scale factor amount, thereby reducing or enlarging the part. (8)

secant. The ratio of the hypotenuse to the adjacent side. It is the reciprocal of the cosine function. (3)

second zero. Machine zero position located a fixed distance from the first zero. It is established by setting parameters. (14)

segment. That part of a straight line included between two points. (3)

servomotors. Variable speed motors that rotate with applied voltage. They drive ball screws and gear mechanisms providing a higher torque output. (1)

setup sheet. Shows how the part is held in a vise or fixture and where the holding device is located on the machine. It may also show where part zero is located, and may provide special instructions to the setup person and operator. (7)(13)

shell end mills. High-speed-steel cutters mounted on special arbors, and ranging in size from 1.25"–6" in diameter. (5)

shell reamers. Different size cartridges, usually 3/4" or larger, that are mounted on either straight or taper-shank arbors. (5)

signed units. Numbers having a value sign of plus (+) for positive or minus (–) for negative. Values to the right of axis Y are positive; those to the left are negative. (2)

simple drill cycle. A drilling procedure that feeds the tool to a specified depth then rapids out of the hole. (10)

sine. The ratio of the opposite side to the hypotenuse. (3)

skim cut. Running a part's edge close to the wire to remove a small amount of material to increase surface finish quality and keep dimensions within tight tolerances. (17)

slurry. A silicon powder added to dielectric fluid (oil) to produce a super matte and highly reflective finish. (17)

spade drills. Used to machine larger holes without requiring a large diameter drill. (5)

spark frequency. All EDMs use a pulse type generator, usually a direct current type that activates and deactivates the current in high frequency pulses. The number of pulses per second is referred to as the spark frequency. (17)

spiral flute taps. Used for tapping blind holes where the material is tough or stringy and chip removal is vital. (5)

spiral pointed taps. Used mostly for through holes, but can be used on blind holes with sufficient chip space at the bottom of the hole. (5)

spot drills. Short drills used to perform the same function as center drills. They are used to accurately locate the position of a hole prior to drilling. (5)

spotface, counterbore canned cycle. Similar to drill, ream canned cycle, except the tool pauses at the bottom of the hole while the spindle rotates. This is done so that tool pressure is relieved. (10)

spring pass. A final threading pass to remove any material remaining at the bottom of the thread due to the flex or spring of the material or tool. (15)

square. A quadrilateral with four equal sides and four right (90°) angles. (3)

Statistical process control (SPC). A tool often used within a TQM system. It involves defect prevention using statistical methods to control a process. (18)

Stellite. Cast alloys made up of approximately 50% cobalt, 30% chromium, 18% tungsten, and 2% carbon. (5)

step blocks. These blocks serve as support for clamps and for clamp height adjustment. (4)

step drills. Drills that provide multiple drilling operations in one tool. These tools are specially designed and sharpened for specific hole configurations. Also called *Subland drills* (5)

stepper motors. Motors that convert an electrical pulse provided by an MCU into a finite rotational step. (1)

straight angle. Angle that is exactly 180°, or a straight line. (3)

subplates. Flat, ground plates with accurately located tapped and reamed holes. Subplates provide a means for locating and fastening workpieces. (4)

subprograms. Programs called by another program called the main program. A machining operation that has to be repeated on the same job can be used as a subprogram. Rather than rewriting the code for the operation each time it is performed, a subprogram is written and can be called within a main program as many times as needed. (16)

supplementary angles. Two angles that equal 180°, or a straight line. (3)

support jacks. This holding accessory is adjustable to provide support under a workpiece to prevent distortion caused by clamping. (4)

S-word. In a program, the letter S followed by a number value denotes the exact desired rpm. (4)

T

tailstock. CNC turning center component. It supports long and heavy work, and can be programmed to function with commands within a program or can be manually operated. (11)

tangent. A line contacting a circle at one point. (3)

tapping canned cycle. This cycle consists of a rapid move to a location, a rapid move to the R-plane, feeding down to depth, reversing the spindle direction, and feeding back to R-plane or initial plane. (10)

tapping. A machining process that produces threaded holes. (5)(11)

T-bolts. Special bolts that engage worktable T-slots, and are used with straps and clamps. (4)

thread milling. Used on external and internal large threads, usually using a thread milling cutter. One machining pass is performed to form the thread. X and Y circular motion forms the thread diameter while Z axis motion forms the pitch of the thread. (8)

threading. Forming internal or external helical grooves. (11)

three-jaw chuck. CNC turning center component with jaws machined to fit snugly around the workpiece. (11)

tool changer. Loads and unloads tools to the machine spindle in response to a programmed command given by the machine control unit. (1)

tool length offsets. Values assigned to tools to allow for the differences in tool lengths. (4)

tool library. A catalog list of all tooling available for each machine in the shop. (7)

tool list. Shows the tools used in the program and usually lists them in the sequence in which they are used. (7)

tool nose radius (TNR) compensation. Used to correct the size and shape of the workpiece when angles and arcs need to be cut on a part. Tool nose radius (TNR) values commonly used are 0.0156″, 0.0312″, 0.0469″, and 0.0625″. (14)

tool presetters. Devices that automatically measure tools held in a turret. (11)

tool radius offsets. Used to designate the size of the cutter being used in a program. A two-digit D word is used in a program to designate the tool offset. (4)

toolholders. Used to secure cutting tools, such as drills, reamers, taps, end mills, face mills, and boring tools. (5)

top clamping. Method using a clamp and clamp screw to fasten the insert to the holder. (12)

Total quality management (TQM). An approach to organizational management that seeks to improve the quality of products and services through ongoing modifications. These modifications are in direct response to continuous feedback. (18)

transversal. A line that intersects two or more lines. (3)

triangle. A three-sided polygon. (3)

trigonometric functions. Numbers called sine, cosine, tangent, cotangent, secant, and cosecant. (3)

trigonometry. The area of mathematics that deals with the relationship between the sides and angles of a triangle. (3)

turning. Reducing the diameter of the workpiece. (11)

turning centers. CNC machines that perform operations including facing, turning, drilling, reaming, grooving, parting, threading, boring, and tapping. Turning centers include the headstock, tailstock, bed, spindle, toolholder (turret), and carriage (slide). (11)

turret. The most common toolholder. It can have an 8-, 10-, or 12-tool capacity. Tools are mounted either upright or upside down. Most turrets rotate in either direction taking the shortest path to select the desired tool. (4)(11)

twist drills. The most common type of drill, available with either straight or tapered shanks. Twist drills may be high-speed steel, carbide, carbide-tipped, or have carbide inserts. (5)

V

variable block. Address format that allows a block of information (instructions) to be any length. (8)

variable data. A quality characteristic that can be measured on a continuous scale. (18)

V-block. A block with a V-shaped notch. Generally used in pairs to support cylindrical workpieces. The blocks are fastened down with clamps or straps, or could be held in a vise. (4)

vector. A line drawn from the origin to the desired point. (2)

vertex. Point or origin where two lines meet. (3)

vertical machining center. Vertical machining centers are similar to vertical milling machines, but use computer numerical control positioning and automatic tool changers to produce complex machine parts, usually in a single setup. The machine spindle is in a vertical position, perpendicular to the machine table. (4)

W

wear offset. Offset used to adjust part size. (14)

Weldon shank. A straight shank with a flat for securing with a lock screw. (6)

word address format. This is the style that programming codes are constructed. (8)

words. A combination of alpha characters (letters) and numerical data. (8)

X

X axis. Horizontal line in the Cartesian coordinate system. When referring to turning centers, the vertical axis is the X axis. (2)

Y

Y axis. Vertical line in the Cartesian coordinate system. (2)

Z

Z axis. When referring to turning centers, the horizontal axis is labeled Z. (2)

zero point. The point at which two axes intersect. Also called *datum* or *origin* (2)

zero return position. A very accurate position along each of the machine's axes that is set by the machine tool builder. Upon machine startup, the machine axes must be sent to the zero return position before operating the machine. (2)

Index